T0351316

Adaptation and Well-Being
Social Allostasis

Recently, an interest in our understanding of well-being within the context of competition and cooperation has re-emerged within the biological and neural sciences. Given that we are social animals, our well-being is tightly linked to interactions with others. Prosocial behavior establishes and sustains human contact, contributing to well-being. *Adaptation and Well-Being* is about the evolution and biological importance of social contact. Social sensibility is an essential feature of our central nervous systems, and what has evolved are elaborate behavioral ways in which to sustain and maintain the physiological and endocrine systems that underlie behavioral adaptations. Writing for his fellow academics, and with chapters on evolutionary aspects, chemical messengers and social neuroendocrinology among others, Jay Schulkin explores this fascinating field of behavioral neuroscience.

DR. JAY SCHULKIN is currently a Research Professor in the Department of Neuroscience at Georgetown University, as well as a member of the Center for the Brain Basis of Cognition at Georgetown. His research investigates the neuroendocrine basis of behavior and his current interests include the evolution of information molecules, such as CRH, oxytocin, behavioral adaptation and the brain.

Adaptation
and Well-Being

Social Allostasis

JAY SCHULKIN
Department of Neuroscience,
Center for the Brain Basis
of Cognition,
Georgetown University,
School of Medicine

CAMBRIDGE
UNIVERSITY PRESS

CAMBRIDGE
UNIVERSITY PRESS

University Printing House, Cambridge CB2 8BS, United Kingdom

One Liberty Plaza, 20th Floor, New York, NY 10006, USA

477 Williamstown Road, Port Melbourne, VIC 3207, Australia

314-321, 3rd Floor, Plot 3, Splendor Forum, Jasola District Centre, New Delhi - 110025, India

103 Penang Road, #05-06/07, Visioncrest Commercial, Singapore 238467

Cambridge University Press is part of the University of Cambridge.

It furthers the University's mission by disseminating knowledge in the pursuit of education, learning and research at the highest international levels of excellence.

www.cambridge.org
Information on this title: www.cambridge.org/9780521509923

© Jay Schulkin 2011

First published 2011

A catalogue record for this publication is available from the British Library

Library of Congress Cataloging in Publication data
Schulkin, Jay.
   Adaptation and well-being : social allostasis / Jay Schulkin.
     p. cm.
   Includes bibliographical references and index.
   ISBN 978-0-521-50992-3 (hardback)
   1. Psychoneuroendocrinology. 2. Sociobiology. 3. Brain – Evolution.
   4. Adaptation (Physiology) 5. Allostasis. 6. Well-being. I. Title.
   QP356.45.S377 2011
   612′.022–dc22   2011002439

ISBN 978-0-521-50992-3 Hardback

Contents

Preface

This is a book on expanding regulatory concepts in cognitive/physiological neuroscience to the context of social adaptation and evolution.

Our evolutionary history is rich in cephalic expansion and innervation into more and more regions of the body, and the corresponding regulation of those regions. The evolutionary history reflects both regulation of the internal and the social milieu.

In recent years, an interest in understanding something about well-being within the context of competition and cooperation has re-emerged within the biological and neural sciences. Darwin understood that prosocial inclinations are built into cephalic regulation, and that set the stage for investigating the ways in which social adaptive mechanisms are involved in establishing and maintaining well-being.

In a previous book, *Rethinking Homeostasis*, I discussed the concept of allostasis from the point of view of regulation of the internal milieu. In this book, I extend the concept of allostasis to the interaction of the individual with the social environment and its influence on regulation of the internal milieu. The book is grounded in an evolutionary perspective with regard to cephalic functioning – specifically to cephalic responses in managing external resources while maintaining internal viability.

An important recent trend in the neurosciences has proven fruitful and warranted: the burgeoning field of social neuroscience. This new field places an emphasis on the biological aspects of social science, including the hormonal regulation of adaptive behaviors as can be seen in the effects of hormones on the brain in generating behavioral responses that serve regulatory needs, including those knotted to the social milieu. The nice thing about this science is the confluence of disparate disciplines merging into a common context for discussion – a context for inquiry.

This book tells a biological story about the importance of human social contact, our well-being linked to significant relationships replete with meaning. Our viability is locked in the social milieu that goes from short- and longer-term considerations, from individual gratification to significant social contact. Given that we are social animals, our well-being is tightly linked to our interactions with others. As such, the discussion will focus on the ways in which social interactions are related to short- and longer-term adaptations.

As always I am grateful for my family and friends. I thank my colleagues for the diverse ways in which we interact; the lifeblood of the mind are the meaningful interactions among us. And I thank my two graduate students in particular, Meaghan Leddy and Britta Anderson.

I apologize for those left out. This book started out much larger, but the editorial suggestion was that it be reduced significantly. Given my personal experience and interest with the amygdala and neuropeptides (e.g. CRH, oxytocin), I have emphasized these topics and interactions with colleagues. For those worthy experiments and individuals not acknowledged, I apologize. The field is large and this book is but a snapshot. Further references and citations may be found at the website for this book, www.cambridge.org/9780521509923.

A visit to Gonville and Caius College, Cambridge University, in June of 2008 was a joy; thank you Joe Herbert and James Fitzsimons.

This book is dedicated to David Jacobowitz, Michael Nitabach and Michael Power: three beautiful scientists.

Introduction

As Aristotle noted long ago, we are, by nature, social animals. Human evolution has increased the importance of social knowledge and social context. Prosocial behavior underlies the moral sensibility that pervades human experience resulting in significant human contact.

Darwin emphasized a fundamental prosocial feature of us, essential for moral judgment. He asserted, "... any animal whatever, endowed with a well-marked social instinct, the parental and filial affections being here included would inevitably acquire a moral sense of conscience, as soon as its intellectual power had become as well or nearly as well developed as in man" (1874: 95). However, in *Descent of Man*, Darwin also noted the "the fewness and the comparative simplicity of the instincts in the higher animals are remarkable in contrast with those of the lower animals" (1874: 65).

Adapting to the social milieu is a fundamental feature of our species. Darwin, like others before and since, understood that we are social animals. What has emerged in *Homo sapiens* has been an elaboration of social contact, the expansion of individual responsibility manifested in specific types of the division of labor in the service of group safety and human well-being and productivity. There has also been a technical expansion resulting in the development of a diverse supply of cognitive resources, including cognitive resources which pitted, at times, deception against social cooperation as conflicting motivations (Dunbar and Shultz, 2007).

We are a vulnerable species; our ontogeny is long and labored and greatly dependent on others. We look to others to gain that important ladder into the social milieu. The long dependency on others is a fundamental feature of our species. The social knowledge we gather in ontogeny represents a critical part of our armament for gaining a foothold in the larger social world; a world in which recognizing others' intentions (e.g., Jaspers, 1913) and gathering practical knowledge are critical. In other words, we come into life prepared to interpret our surroundings as defined by the social milieu, and the degree to which we succeed in this task determines to a great extent our success in coping, achieving and thriving. The fact that we come prepared to recognize others and learn from their experiences is thus a fundamental social behavioral adaptation.

Moreover, what is distinctive about us, although our species is not alone, is the degree to which we share and participate toward common ends; shared intentions linked to the considerations of others is one of our most important cognitive adaptations. We look at others; it is not surprising that vision, shared visual space, and recognition (that we are all looking at the same objects) would come to be important cognitive resources. But it is not simply a cognitive detached event; it is affectively rich, reassuring and rewarding. The motivation to form meaningful contacts is essential for development and for life.

Depicted below are some common themes in our cognitive development, particularly that of social development, in Table I.1 (adapted from Tomasello *et al.*, 1993).

GENES, BRAINS AND BEHAVIOR

Changes in the internal milieu (e.g. hormonal secretion) have been long noted. Human prenatal and postnatal development is long and varied, tied as it is to the necessity of the acquisition of a huge body of knowledge during the early period. There is profound change in brain morphology and development during the protracted neonatal period in varying degrees in most mammals (McCarthy, 2008; O'Doherty *et al.*, 2004). Hormones, such as oxytocin, vital for parturition, for the birth process, lactation, and social attachment, figure throughout the gestational and developmental periods.

Oxytocin is produced in the placenta and the brain and the pituitary gland; the fetus is awash in information molecules as it floats in the amniotic fluid within the safety of its mother. Learning about safety and comfort in the new and harsh external world begins in the immediate postnatal period, as the neonate wails and is quickly allowed to suckle.

When one is trying to understand the behavioral response, the brain must be considered, as well as how oxytocin is regulated in the brain. Steroids such as the gonadal hormones or the adrenal

Table I.1. *Infancy: understanding others as intentional (Tomasello et al., 1993)*

1. Following attention and behavior of others
2. Directing attention and behavior of others
3. Symbolic play with objects
Early childhood: language
1. Linguistic symbols and predication
2. Event categories
3. Narratives
Childhood: multiple perspectives and representational re-descriptions
1. Theory of mind
2. Concrete operations
3. Representational re-description

hormones impact the degree of neuropeptide expression (e.g. oxytocin) that underlies behavioral approach and avoidance systems (see below). The genes that underlie, for instance, the production of the peptide oxytocin, in combination with external circumstance, contribute to developmental outcome (e.g. Carter, 2007). They promote social attachment, or approach behaviors (see Chapter 5). One should note that too much is at stake to put everything into oxytocin for the diverse social behaviors of approach that we note in our everyday lives (Chapters 4 and 5).

Disorders such as autism, for example, appear to be associated with an aberration in the motivation for social contact, expressed as extreme social withdrawal (Baron-Cohen, 1995/2000). It is noteworthy that autism, a devolution in social competence, is more heavily expressed in males, and perhaps is linked to the expression of gonadal steroid regulation or sex-specific organizational changes in the brain (Baron-Cohen *et al.*, 2004), perhaps via oxytocin expression (Hollander, 2003).

The social isolation linked to autism is maddening, perhaps provoking a propounded sense of fear that goes along with the isolation. Isolation of this form is not adaptive and in the case of autism a developmental disability has a strong link to brain function and the genes that regulate diverse information molecules, including that of oxytocin (Carter, 2007).

Perhaps it is not surprising that a basic human form of social contact, namely eye contact, would be so severely degraded in autism (Baron-Cohen, 1995/2000). We all know that the sense of security that we derive when we form solid social bonds is quite comforting; this is

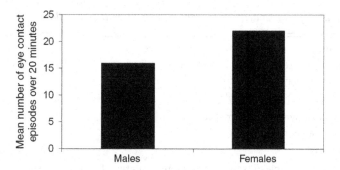

Figure I.1   Females make more eye contact than males (adapted from Baron-Cohen *et al.*, 2004)

impaired in autism. And autism is more pronounced in males (Baron-Cohen *et al.*, 2004). Moreover, quite early in development, male and female children are quite different with regard to eye contact; girls make contact quite a bit more. A diverse set of genes, including those which affect oxytocin, are linked to diverse forms of social devolution (Hollander, 2003).

A diverse set of neuroendocrine systems bound to cognitive capacity is encoded to ensure meaningful contact in development, and throughout our lives; devolution of function (e.g. autism) is the diminishment of this important capacity. Furthermore, our evolution favored social contact: group formation in which diverse cognitive skills facilitate the formation of social bonds. Central peptide systems (e.g. oxytocin, vasopressin), which will be described in subsequent chapters, are linked to social adaptation and function (see also Chapters 5–7).

From the point of view of the brain, two kinds of regulation are taking place: the regulation of the internal milieu, and adaptation and regulation of the social milieu. Both require a brain with diverse physiological and behavioral regulatory systems. In both cases, anticipatory mechanisms underlie adaptation both within the individual and within the social milieu. The greater the degree of social contact and social organization experienced by a human, the greater the trend for cortical expansion (Figure I.2, Dunbar and Shultz, 2007).

Nature selected physiological cognitive systems oriented to social systems. Their evolution and expression underlie the diverse forms of complicated social assessments; group size, for instance, is correlated with cortical expansion (Dunbar and Shultz, 2007). Consider the

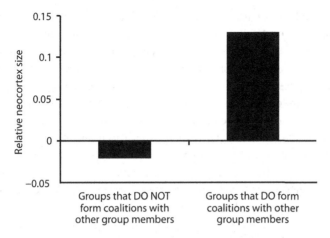

Figure I.2   Social contact in primates is consistently linked to neocortical expression and size (adapted from Dunbar and Shultz, 2007).

complex social relationships of primates the hierarchy, distribution of food resources, shelter protection, dominance, and comfort through co-alliances. Such systems are quite varied and all involve cephalic innervations and expression.

However, social behavior is just one element in our evolutionary ascent, along with an evolved motor system in which subcortical brain regions are pregnant with cognitive capacities. Structural changes in our visual system and bi-pedalism are most definitely core features in our evolutionary ascent.

ALLOSTASIS

The concept of "allostasis" was introduced to take account of the physiology of change and adaptation to diverse circumstances, and to the behavioral and physiological anticipation of future events (Sterling and Eyer, 1988; Sterling, 2004). The concept has been centered in cephalic anticipatory regulation of the internal milieu in the context of the social framework: changing circumstances, adapting to change, and features that are not fully grasped in traditional accounts of homeostatic regulation (such as the maintenance of glucose levels at one level) of the internal milieu (Schulkin, 2003).

Walter Cannon, a student of William James at Harvard at the end of the nineteenth century, popularized the concept of homeostasis; he had an evolving sense of this concept. It was far from a simple one dimensional conception, namely rigidly guarding a simple set point (a body's natural state of balance). Despite Cannon's dispute with his mentor, William James, with regard to James' peripheralist conception of the emotions (e.g. knowing that one is afraid because one's heart is beating fast), Cannon did not emphasize the anticipatory. Rather, he emphasized adaptation and the exhaustion factor that can occur under extreme duress (e.g. voodoo death-shock death); it was just not part of his scientific lexicon, nor was a consideration of behavioral adaptation in the regulation of the internal milieu – that was left for others and for those who were influenced by Cannon (e.g. Richter, 1943) – though Cannon, and later his students, revealed that different parts of the brain were obviously essential for the generation of physiological and behavioral features of adaptation to duress.

In fact, to take account of the social regulation of the internal milieu in addition to the anticipatory mechanisms in maintaining internal viability, another sort of regulatory concept was necessary. In this regard, allostasis is, in part, the process by which an organism achieves internal viability through bodily change of state, which comprises both behavioral and physiological processes that maintain internal parameters within the limits essential for life.

"Allo" means change, whereas "stasis" means stable/same. But the social milieu changes and within this we invoke both cognitive and physiological resources to promote continuity, stability and predicted outcomes – the "allo." Part of "stasis" is about adapting to change to achieve the goal of stability in the face of uncertain circumstances, something all of us know about early in life and continue to experience throughout life; it is not a very abstract concept, it is up close and personal and pervades our experiences. Thus, allostasis is about adapting to change and anticipating the need for change, so as to restore the base state within a new physiological or environmental context (Sterling and Eyer, 1988).

Allostasis developed in response to a need to conceptualize adaptation to change in a way that took account of all the exigencies of the environment and changing circumstance (Wingfield, 2004). The emphasis is on how we achieve internal viability in adapting to changing circumstances within parameters essential for life processes; chronic overactivity of regulatory systems render one vulnerable to pathophysiology.

Allostasis emphasizes regulation that is an adaptation to change; not just in reaction to it, but in anticipation of it (Bauman, 2000). Unpredictable events are a constant feature of the life cycle for most animals, and the need for stability and consistency are a constant characteristic of most animals, within both the physiological and social domains, allostasis is a means of achieving stability in the face of unpredictability.

Life cycles for many species are both predictable and unpredictable. Seasonal changes are predictable, as are light/dark cycles, but unpredictable events obviously pervade the seasonal changes. They include climate changes, such as severe droughts, food shortages or disruption of social status.

Disruption of social status, like most unpredictable events, impacts cortisol levels. Levels of cortisol, for instance, by itself are not a measure of dominance, but of the way in which dominance is achieved (Cavigelli *et al.*, 2003). One recurrent theme is that social support is predictive of cortisol levels, mitigating diverse events which would elevate it in combating adversity (Abbott *et al.*, 2002). Consider what it is like for a wolf pup under the care of the alpha female, who heads the group; when she is killed, the offspring is vulnerable. How vulnerable depends upon social alliances, the age and strength of the offspring, and the terrain and social milieu.

Many different cognitive and physiological systems are involved in anticipating future events, including biological events that factor significantly for survival. The cephalic expansion of cognitive capacity is reflected in these biological systems. Homeostasis has been linked to reflexive reactions, not cognitive anticipation of events and cephalic innervation and regulation of peripheral physiological adaptation. Reflective reactions underlie the immediate sizing up of a social event, through the eye contact and bodily postures, through forms of encoding complex and not so complex social relationships. Think about how fast we are as a species, although we are sometimes utterly mistaken, in our assessment of possible threat or possible social comfort.

One feature of the human brain is the presence of multiple mechanisms that provide the potential for cephalic anticipatory adaptation – that is, the brain constantly attempts to anticipate future events. This is paramount in larger cast social orders, and adaptation is achieved, in part, by cephalic regulation of behavioral and physiological systems in the expression of longer-term adaptation (Sterling and Eyer, 1988).

Traditional conceptions of regulation have typically (though not always) emphasized homeostatic short-term regulation, with further focus on set point stability and short-term contemporaneous adaptation (Cannon, 1916). The expansion of cortical function, in social groups of diverse complexity, entails longer-term regulation.

My emphasis is also on cephalic changes as a result of behavioral adaptation, involving physiology through social relationships, building an environment, use of tools, etc. And this is fundamental to the concept of allostasis. In addition, the concept of allostasis and allostatic overload, pushing the system beyond its adaptive capability, is designed to account for vulnerability to disease by taking into account variations in individual experiences and genetic makeup, and to determine how a lifetime of short- and longer-term adaptation affects wear on the body. Natural aging is one such impact; social inequities and nutritional strains are others. Biology is designed to promote both short- and longer-term viability.

An example is in order: one consequence of high levels of cortisol for long periods of time is vulnerability for memory deterioration via hippocampal damage (McEwen, 1998) and the chronic overactivation of diverse regulatory systems. The hippocampus, an anatomically beautiful structure, is importantly involved in memory, and individuals vulnerable to defects in memory function appear to be chronically compromised by elevated cortisol levels over extended periods of time, a situation that results in neuronal degradation.

In other words, neurons essential for memory are degraded, but the degradation may not be permanent and linked to compensatory responses; diverse growth factors in the brain are regulated by diverse steroid hormones, including estrogen, testosterone and cortisol. The steroids regulate the genes that underlie the expression of diverse growth factors essential for neuronal and other bodily tissue – that is, keeping the tissue afloat and functional – and under diverse conditions these factors are differentially regulated by these various steroids.

On the adaptive side, cortisol is important in the conversion of short- to longer-term memory (McGaugh, 2000). Experiments demonstrate that cortisol, perhaps by facilitating norepinephrine, influences memory formation; it is a fundamental neurotransmitter in the brain which contributes to the consolidation and induction of short- to long-term memory – broadcasting the memory to diverse regions of the brain. In other words, cortisol impacts regions of the amygdala, which is vital for memory formation.

Several examples come to mind: the memory of John Kennedy or Martin Luther King being shot; the Taliban bombing of ancient monumental figures of Buddha; and the bombing of the World Trade Center. The consolidation from short- to longer-term memory is enhanced by the broadcast of this information by the effects of cortisol in the brain (McGaugh, 2000).

I spent years looking at the World Trade Center from my New York apartment. Though living in Washington at the time of the destruction of these buildings by suicide bombers, no doubt the induction of and regulation of diverse neurotransmitter and neuropeptide systems have facilitated the stability of that memory.

Put slightly differently, the induction of the neurotransmitters related to adrenaline in the brain, and the regulation of their expression in diverse regions of the brain (e.g. amygdala, hippocampus, regions of the neocortex), then broadcasts the memory event to diverse regions of the brain that underlie memory. Importantly, cortisol has diverse effects on end organ systems in regulating glycogenesis and whole body regulation, and one such form of regulation involves central neurotransmitters and the facilitation of memory formation. Memory is the key to a consideration to the self; no memòry no self, is one suggestion.

There is an important social adaptation designed to reduce elevated levels of cortisol back down to normal. Social attachment – the formation of core safe (or relatively safe) attachments – provides feedback on the expression of steroid levels, such as cortisol, and the regulation of the internal milieu.

The emphasis, however, is on social competence (which would include cooperative behavior in tool use) but also, since we are social animals, an evolved set of anticipatory mechanisms that facilitated the regulation of the internal milieu amidst an expanding social milieu. And these behavioral adaptations impact the brain directly – evolution selected for behavioral adaptations that feed back directly onto cephalic function.

REGULATION

Regulation means many things. It surely means anticipation of events, for instance, anticipation and absorption of food resources vital for bodily functions. Hoarding resources in anticipation of metabolic needs and in anticipation of seasonal needs is not an uncommon behavior (Wingfield, 2004).

Cephalic organization underlies behavioral adaptation. Consider one example: Curt Richter, a psychobiologist at Johns Hopkins, demonstrated in his laboratory a well-known ethological fact about many animals in nature. The laboratory rat constructs a nest to keep warm as a form of thermal regulation in anticipation of temperature demands. It is a fairly common occurrence in humans as well; tool use combined

Table I.2. *Cephalic function and ingestive behavior (Power and Schulkin, 2009)*

Cephalic/Physiological	Organ	Function
Salivation	Mouth	Lubrication, begin digestion
Gastric acid secretion	Stomach	Hydrolysis of food
CCK	Small Intestine	Short-term feeding
Insulin	Pancreas	Regulates glucose and fat storage
Leptin	Adipose tissue, stomach	Reduces appetite
Ghrelin	Stomach	Stimulates appetite, fat absorption

with anticipatory activity (e.g. more than reaction to maintaining one set point) is a core feature cephalic expansion into the regulation of the internal and social milieu.

Cephalic machinations integrate our internal physiology by behavioral regulation (Richter, 1943), such as the rich innervation of peripheral sites or the activation of peripheral sites by vagal efferent projections into sites along the digestive tract.

Importantly, cortical sites (e.g. the frontal cortex) project to these brainstem sites; expansion of cortical function into larger forms of visceral control is a core theme throughout this book, and reveal a diverse anticipatory regulation of the internal milieu.

Diverse anticipatory systems are expressed as we are about to ingest a food resource, many of which are the rich array of information molecules expressed in the brain and peripheral nervous system in the absorption and utilization of food resources (Swanson, 2000). A partial list is depicted in Table I.2.

Peptide hormones in this case are expressed in both the peripheral and central nervous system, and, of course, they work in the context of the larger physiological and behavioral control of food ingestion (Herbert, 1993).

Pavlov, (1927) described a "cephalic phase," an anticipatory response to food ingestion; insulin is secreted in anticipation of food. These events are cephalic; the context of food passing through the oral cavity facilities the anticipation of utilizing and distributing the vital food resources, and, therefore, in this context insulin is secreted in advance of absorbing the nutrients (Powley, 1977).

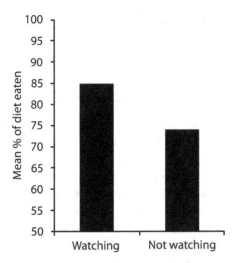

Figure I.3   This graph depicts an experiment in which an omnivorous rat is learning what to eat in part by watching a con-specific discover and eat safe foods. Mean percentage of demonstrator's diet eaten by observers assigned to uncertain (dark) and certain (light) conditions (adapted from Galef and Whiskin, 2000).

In addition, food resources that have been associated with visceral illness decrease food intake and impact cephalic anticipatory responses; rats show degraded levels of insulin when presented the same food resource that has rendered them with visceral distress (Berridge *et al.*, 1981). Learning impacts the secretion of information molecules, such as insulin (Woods *et al.*, 1970).

In many species what to approach or avoid is socially facilitated; watching con-specifics ingest food sources provides an orientation of what to ingest and what to avoid (Galef and Whiskin, 2000, Figure I.3), which contributes in the regulation of internal milieu.

The neuropeptides and neurotransmitters are regulated directly by many behavioral adaptations (Carter *et al.*, 1997/1999). The sight of a predator (a dog, for instance) elicits an immediate release of diverse central information molecules (bombesin, corticotrophin releasing hormone); CRH measured by microdialysis in sheep is but one example (Cook, 2002). CRH in the amygdala is released under diverse conditions, particular under novelty. The ability for the sheep to remove themselves from the vicinity of the dog, results in a rapid diminution of CRH; behavior serves to regulate central CRH expression. The instrumental behaviors tied to successful responses, avoiding an aversive noxious event has an immediate impact on CRH expression. CRH, amongst other things, is a neuropeptide tied to attending to noxious and fear related events.

SOCIAL AND BEHAVIORAL REGULATION OF THE INTERNAL MILIEU

Regulation of the internal milieu is linked to the social environment, a connection that succeeds or fails in the balance between cooperation and competition for resources. What appears to have evolved in *H. sapiens* is a "cognitive penchant" for long-term considerations derived from social cooperation and social knowledge (Darwin, 1859/1958).

This "cognitive penchant" coexists with the expression of individual differences in developmental vulnerabilities. Social shyness, or fear of public disclosure, is a painful and psychological and physiological expensive event; it is no fun being excessively shy and then being put in a position of social performance.

For instance, vulnerabilities to social phobia and other fear-related patterns are linked to social context and social habit as well as to genetic makeup. There is some evidence that shy children have genetic differences in the level of the CRH and serotonin gene receptors (Wagner *et al.*, 2006). Young children who are more fearful of unfamiliar social events tend to secrete more cortisol from the adrenal gland, and when later tested have greater amygdala activation to unfamiliar social scenes (Schwartz, *et al.*, 2008). It is not the elevation of cortisol by itself but the induction of corticotrophin releasing hormone gene expression in regions of the brain that contribute to fearful social wariness (Schulkin *et al.*, 2005).

One important mediator for both physiological and behavioral adaptation is cortisol – a steroid secreted by the adrenal gland. Glucocorticoids are primarily related to the organization of glucose metabolism, something in which all cells are implicated. They participate with metabolic hormones (e.g. insulin) in the regulation of metabolism and in the generation of searching and feeding behavior, in part by the activation of diverse neurotransmitters and neuropeptides. In fact, diverse forms of appetitive behaviors are associated with elevated levels of cortisol. They are importantly linked to diverse forms of foraging behaviors, and anticipation of rewards (Wingfield, 2004).

But cortisol function is broad in regulatory behavioral and physiological regulation; moreover, glucocorticoids vary their effects in seasonal variation, in droughts and in predictability for the allocation of food resources and other metabolic requirements. Glucocorticoids are not a one-dimensional entity. They are multidimensional in action and effect, and thus fundamental to biological and social adaptation. Their effects are both short-term adaptations and long-term detriment, which are depicted in Table I.3.

Thus, glucocorticoids, for instance, promote foraging behavior in many different animals under various conditions, and facilitate life processes essential for successful reproduction, energy metabolism and attention to external events. The important point in understanding regulatory physiology promoting viability over time is the adaptive nature of actions of different hormones depending upon the circumstances both internal and relative to niche and circumstance.

Cortisol, long referred to as the hormone of "stress," is mischaracterized when understood in this way. It is the hormone of energy metabolism and adaptation through which different end organ systems are mobilized or restrained for action. An important consideration in the successful aging process, a cumulative biological marker, is the regulation of cortisol, namely the ability to secrete more of it when necessary and terminate the elevated levels when unnecessary. These are ameliorated or not by social context, genetics and Fortuna.

The secretion of cortisol is an adaptive response serving physiology and behavior. The continued secretion of cortisol without relief has consequences though; bone demineralization, compromised immune system and shifts in metabolism are the result of secretion of cortisol (Sapolsky, 1992). One important adaptation is regulation of the internal milieu – that is, the secretion of cortisol – through manipulations in the social context.

For example, cortisol tends to be elevated in troops of macaques under conditions of uncertainty, including uncertainty of social control (Sapolsky, 1992). Perception of where an individual baboon is in the hierarchy of its social group influences the internal milieu, particularly cortisol levels. In addition, social perception, facilitating social approach or avoidance, results in the behavioral regulation of cortisol (Cheney and Seyfarth, 2007).

However, it is not axiomatic that cortisol is elevated in the downtrodden, insecure members of primate groups. Diverse behavioral correlations can result in elevated levels of cortisol in many contexts (Abbott et al., 2002). It is not the case that elevated cortisol levels equals adversity. The important point about cortisol is not whether it is elevated in the dommiant or the socially less fortunate

Table I.3. *Glucocorticoids and diverse forms of behavioral adaptation (adapted from Goymann and Wingfield, 2004)*

Species	Sex	Source	Type of glucocorticoid and technique	Factors other than social status affecting GC levels
Harris's hawk, *Parabuteo unicinctus*	Female	Plasma	Corticosterone (RIA)	Breeding season
	Male			Breeding season, age
Florida scrub jay, *Aphelocoma coerulescens*	Female	Plasma	Corticosterone (RIA)	
	Male			
White-browed sparrow weaver, *Plocepasser mahali*	Female	Plasma	Corticosterone (RIA)	Breeding season
	Male			
African elephant, *Loxodonta Africana*	Female	Faeces	GC metabolite (EIA)	Group size
Alpine marmot, *Marmota marmota*	Female	Faeces	Cortisol (EIA)	Breeding season
	Male		Cortisol (RIA)	Reproductive status
Spotted hyaena, *Crocuta crocuta*	Female	Faeces	GC metabolite (EIA)	Reproductive status, commuting
	Male			Proportion of males per clan or clan size
African wild dog, *Lycaon pictus*	Female	Faeces	GC metabolite (RIA)	
	Male			
Wolf, *Canis lupus*	Female	Faeces	GC metabolite (RIA)	Breeding season
	Male			Breeding season
Ringtailed lemur, *Lemur catta*	Female	Faeces	GC metabolite (RIA)	
Savannah baboon, *Papio anubis*	Male	Plasma	Cortisol (RIA)	Personality, social support
Mountain gorilla, *Gorilla beringee*	Male	Urine	Cortisol (RIA)	Time of day, age
Chimpanzee, *Pan troglodytes*	Male	Urine	Cortisol (RIA)	Fruit availability, aggression

group member, but whether there are behavioral strategies for regulating cortisol – turning it off, decreasing and increasing the level when necessary and its costs when not, social status, social comfort being one. In an interesting confluence of two disciplines, the cognitive ethological literature and the endocrine regulatory literature, the social milieu is long known in both to help regulate steroid levels (Sapolsky, 1992).

In humans, like several other mammalian species (e.g. macaques, baboons; Sapolsky, 1992), social contact under conditions of uncertainty can reduce cortisol secretion. For instance, grooming behavior is a fundamental form of social organization. Grooming behavior is involved in sustaining alliances amongst group members, in the formation of new alliances, appeasement, comfort and attachment – "friendships" during conditions of adversity impact, for instance, cortisol levels (Cheney and Seyfarth, 2007, Figure I.4). In other words, complex alliances are kept on track and are modulated by grooming behaviors. This pattern of behavior, because it is linked to tracking of complex hierarchical relationships, represents a highly cognitive adaptation fundamental for group coherence.

Social contact is one way of managing the secretion of cortisol – more positive social contact, more grooming, and therefore less cortisol circulating; forming coalitions is essential for the regulation of the internal milieu, of which cortisol, the molecule of energy metabolism, is one molecule amongst others being regulated by social contact. In a similar way, young children who form social contact and seek

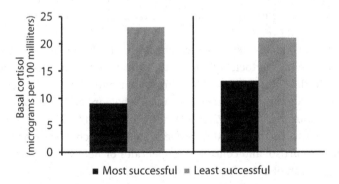

Figure I.4   Cortisol, the adrenal hormone is secreted (or not) under diverse conditions, one of which is uncertainty of social stability (less successful) in the rough life of a baboon troop (adapted from Sapolsky, 1995).

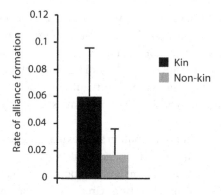

Figure I.5   Social alliances, kin relationships are an important behavioral adaptation of the vervet troop behavior. Rates at which adult females in vervet groups formed alliances with each of their kin and nonkin (adapted from Cheney and Seyfarth, 1990).

out human warmth and security tend to have lower levels of cortisol under diverse conditions (Gunnar and Davis, 2001).

An important adaptation is the formation of social contact, keeping track of alliances, amongst other social events (Cheney and Seyfarth, 2007, Figure I.5)

The social context then feeds back upon the secretion of the steroid. And steroid hormones such as cortisol, as I will make clear in this book, have diverse effects on the brain; they regulate the production of genes that produce neuropeptides (e.g. CRH), which are a common final pathway in promoting the behaviors that underlie approach and avoid, sustain and persevere in the myriad of social and complex behaviors. With coalitions or friendships cortisol is a bit lower in studies of primates and other animals in the wild (Engh *et al.*, 2006, Figure I.6).

For instance, chemical messengers in the brain, such as CRH or vasopressin, oxytocin or prolactin, are regulated by steroid hormones in different regions of the brain that underlie diverse behaviors, including approach and avoidance behaviors and physiological systems (Carter *et al.*, 1997/1999). One recurrent and constant theme in this book is the relationship between steroids, such as estrogen, testosterone, cortisol and central neuropeptides or neurotransmitters in the regulation of behavior.

Another example is the effects of estrogen on oxytocin gene expression. In many species estrogen is known to facilitate central oxytocin secretion and different forms of maternal behavior (Keverne

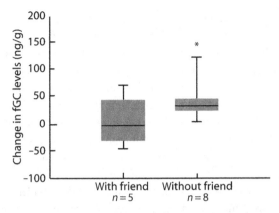

Figure I.6   Change in female glucocorticoid (fGC) levels during a period of infanticidal attacks for lactating females with and without male friends. Reliable alliances are a fundamental feature for longevity; cortisol is higher without strong friendly alliances in this troop of female baboons (Engh, A. L., Beehner, J. C., Bergman, T. J., et al., (2006). Female hierarchy instability, male immigration and infanticide increase glucocorticoid levels in female chacma baboons. *Animal Behavior*, **71**, 1227–1337).

and Curley, 2004). It is the induction of oxytocin gene expression in various regions of the brain that underlies this complex behavior. These areas are limbic regions of the brain, including the hypothalamus, amygdala and bed nucleus of the stria terminalis that underlies diverse forms of social behaviors (Davis *et al.*, 2010).

Of course the social milieu has diverse effects on end organ systems, such as the brain, and the regulation of oxytocin is dependent upon the social milieu. That is, the social milieu helps determine the level of estrogen and the degree of gene expression in the brain; but it is the information molecule oxytocin that is importantly involved in sustaining the behaviors associated with attachment behaviors (Carter *et al.*, 1997/1999; Neuman, 2008).

Again, consider an example: steroid hormones such as cortisol, estrogen, and testosterone maintain genes that regulate peptides and their receptors; these relationships underlie a plethora of adaptive social responses. Withdrawal of these hormones decreases the expression of, for instance, vasopressin (DeVries and Miller, 1998), which is important for diverse forms of social behaviors.

In social relationships, for instance, gonadal steroid induction of vasopressin and territorial markings are well established. One expression of this is urination and marking off territory; vasopressin is also fundamental to water regulation and retention (Denton, 1982). It is interesting that this hormone would be linked to marking territory, defending territory and a form of social regulation. It also applies to information from scent through olfaction transduction mechanisms; there are suggestions that vasopressin/oxytocin at the level of olfactory systems from the peripheral to the central nervous system (outside or inside the cranium) are a fundamental means of social communication; marking space, territory, etc. Another well-known example is the induction of neuropeptides by steroids that facilitate social approach (oxytocin) and social withdrawal (CRH) within one region of the brain – the amygdala. But it will be couched within a social framework (Carter, 2007).

Interestingly, the chimpanzee, our nearest primate relative, may be a case in part, though speculative, in which oxytocin expression and/or CRH expression is different in the Bonobo chimpanzee from either us or its larger cousin. The Bonobo chimpanzee (the smaller variant) is extremely peaceful, and uses sexual alliances to reduce social tension and forms coalitions that are peaceful when contrasted with the larger chimpanzee (DeWaal and Lanting, 1997). The Bonobo is one of those rare species, along with our own, that face each other in bouts of intimacy; one would predict that one gene that is part of the Bonobo's behavior is oxytocin and would be differentially expressed and regulated than its more aggressive cousin. That is, oxytocin should play a larger role in the behavioral regulation of the social milieu in the Bonobo than in *Pan troglodytes*.

MEANINGFUL SOCIAL CONTACT AND HUMAN WELL-BEING

An essential part of our well-being is in the link to others; human flourishing is bound to others. It is within the social milieu in which approach and avoidance behavioral responses for participation with others occurs; our evolutionary machinery sets the stage for this modern and ancient participation with others and for our sense of well-being.

Human well-being is in part a physiological concept, but because we are social animals, it almost always involves our relationship to others, the diverse and essential meaningful experiences we derive from social contact. Friendship, compassion and connectedness to one another are essential for our mental health and our sense of well-being. The value of friendships is something featured in other primates as essential for normal development and long-term viability. As Karl Jaspers, a noted existential psychiatrist noted at the beginning of the 20th century stated: "… we are always being led by ideas to a complex unity of meaningful connections …" (1913/1997: p. 760). The search for meaning underlies human existence, and a great deal of it is social in nature. As the poet Robert Burns put it: "The deities that I adore are social peace and plenty" (from "Song"). An evolved social framework is an essential feature of the central nervous system; and what evolved are elaborate behavioral ways in which to sustain and maintain the physiological/endocrine systems that underlie the behavioral adaptations, which is the subject of the subsequent chapters.

Thus in what follows, Chapter 1 provides a perspective on evolutionary change, including punctuated equilibrium, along with a conception of allostatic physiological changes that sets the context for understanding social adaptation. Chapter 2 is concerned with the cognitive and neural resources for the formation of social contact. Chapter 3 looks at some architectural and functional features of the brain. Chapter 4 depicts the diverse information molecules, their origins, and diverse functional role that underlie cerebral design and behavioral adaptation. Chapter 5 explores social neuroendocrinology that underlies approach and avoidance behavioral expression, an example of allostatic regulation, and biobehavioral adaptation. Chapter 6 provides a perspective on adaptation and the response to diverse incentives that serve viability and devolution, a deterioration of function. In addition, the chapter provides an orientation to understanding an analysis of the reward signals in the brain as they relate to the internal milieu and the external conditions and the vulnerability to allostatic overload. Chapter 7 builds on the origins of what Darwin called prosocial behaviors, a forerunner for ethics; and the concluding chapter is a discussion of well-being in the age of neuroscience.

Each chapter is bound to the concept of allostasis and cephalic machination in both short- and longer-term adaptation. Each chapter will also discuss how the adversity of modern life is a threat to well-being, and why social group interactions may be more important than ever before.

1

Evolutionary Perspectives and Hominoid Expression

INTRODUCTION

A key feature in evolution is how well the behavioral/physiological adaptation works, and how flexible the particular adaptation can be when expanded into diverse contexts. Our evolution reveals a conception, not always accurate, that rigidity is a feature of lower species, whereas flexibility is a feature of primates such as us. Corticalization of function reflects the larger role that social, cognitive and anticipatory regulation of the social and internal milieu play in the organization of behavioral and physiological viability (James, 1890/1952).

Responding to diverse social signals requires a flexible brain with a range of behavioral options. Our evolution is linked to our cognitive competence and one feature of this is our social contact, both cooperative and competitive. Competition often requires cooperative understanding to ensure success, depending upon one's view of evolution and the cultural context in which cooperation and competition are understood. Evolution has selected both and one should be wary about mythologizing one or the other in our conceptual framework. This chapter orients the reader towards several conceptions of evolution and their importance in understanding our cephalic expansion, which underlies our diverse forms of behavioral and physiological adaptations to changing contexts.

EXPLANATIONS IN THE CONTEXT OF EVOLUTION

The great taxonomists, starting from Aristotle, have created long catalogues of natural objects and, in this case, the lineages relating biological taxa and species. Recognition of kinds of objects is a predilection that comes with our cognitive endowment; we are taxonomic animals. Evolution favored this cognitive ability in our species. Perhaps there is a predilection to pick out kinds of objects as they appear in nature and have an inherent interest in their investigation, in which ontogenetic and phylogenetic development and continuity and change figure importantly (Gould, 2002).

Evolutionary explanations have ancient beginnings, but reached a critical mass during the years prior to Charles Darwin and culminated with him and perhaps Mendel at another level of analysis. As Lamarck noted "how great is the antiquity of the terrestrial globe and how small are the ideas of those who attribute to the existence of this globe an age of six-thousand and several hundred years from its origin in time" (cited in Burkhardt, 1975/1997, p. 110, from Lamarck's *Hyrdrogeologic*). All explanations extend back to a conception of evolution, with natural selection as a key concept. Reproductive fitness, sexual selection, variation and speciation are some of the key categories, in addition to more modern terms that take into account genetic drift, as Mayr (1963) noted in the years since Darwin's fundamental and revolutionary insights. However, natural selection is the one all-inclusive concept. One depiction of the key insights from Darwin and others is depicted in Figure 1.1 by Ernst Mayr (1991), one of many expositors who spanned the whole of the twentieth century and wrote many books on Darwin and evolutionary biology. The development of different populations, the adequacy of resources and the genetic variation and natural selection are highlighted.

Evolutionary biologists have realized that selection is more varied, involving the opportunistic spread of genetic material that enlarges our conception of natural selection, but it still remains the

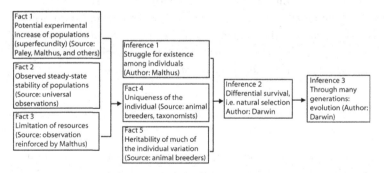

Figure 1.1   Darwin's explanatory model of evolution through natural selection (Mayr, 1991).

key category explaining why functions are made apparent. While selection is a primary feature in evolution, it is not always synonymous with adaptation. Another view of evolutionary change is depicted in Table 1.1. (adapted from Simpson, 1944 in Eldridge, 1985.

Change and adaptation, variation and reproductive successes are the key biological variables, and not the glorified trend towards progressive functions. Evolution via natural selection is about adaptation to fit niches, opportunistic use of specific functions in more varied environments and novel functions of existing structures and molecules. Research in the nineteenth century discovered an evolutionary perspective through which one could understand the rest of biology, as it was noted later in the twentieth century (Dobzhansky, 1962). Darwin is at the heart of this; again, two factors, natural selection and variation in adaptation, are key ingredients.

Conceptions of geological and climatic change have also been at the heart of evolution, and especially the extinction of species that permeated the air that Darwin was breathing intellectually. Lyell in his great book *Principles of Geology* devotes several chapters to both concepts. Geographical variation and species differences had become a core concept as had those of competition and survival, and resource availability and allocation, as Malthus (1798/1970) depicted in his work, *An Essay on the Principle of Population*. At various points in our ecological history there may have been bouts of rapid climatic change that resulted in the extinction of species (e.g. 2.5 million years ago) and perhaps another moment of rapid change from within our own lineage that took place about 100,000 years ago in which there was a greater preponderance of instability or unstable environments. Our evolution is replete with change and stability; the climatic change perhaps facilitated the dispersal of hominoids across diverse regions as an adaptation to climate change (Foley and Lahr, 2004). Speciation by geographical separation is a fundamental mechanism through which wide variation is made manifest. Variation and speciation are the lifeblood of evolutionary change (Darwin, 1859/1958), amid pockets of abrupt change and longer-term viability (Gould, 2002).

Variation is at the heart of evolution, too, in addition to extinction and the formation of speciation by geographical isolation. The same species separated by space and habitat can produce two very different adaptations; for example a land-dwelling and sea-adaptive iguana; geographic differences coupled with habitat select for diverse forms of adaptation.

Table 1.1. *One depiction of evolutionary change (adapted from Simpson, 1944 in Eldridge, 1985)*

Mode	Speciation	Phyletic evolution	Quantum evolution
Typical taxonomic level	Low: subspecies, species, genera, etc.	Middle: genera, subfamilies, families, etc.	High: families, suborders, orders, etc.
Relation to adaptive grid	Subzonal	Zonal	Interzonal
Adaptive type	Local adaptation and random segregation	Postadaptation and secular adaptation (little adaptive or random change)	Preadaptation (often preceded by inadaptive change)
Direction	Shifting, often essentially reversible	Commonly linear as a broad average, or following a long shifting path	More rigidly linear but relatively short in time
Typical pattern	Multiple branching and anastomosis	Trend with long-range modal shifts among bundles of multiple isolated strands, often forked	Sudden sharp shift from one position to another
Stability	Series of temporary equilibria, with great flexibility in minor adjustments	Whole system shifting in essentially continuous equilibrium	Radical or relative instability with the system shifting toward an equilibrium not yet reached.
Variability	May be temporarily depleted and periodically restored	Nearly constant in level; most new variants eliminated	May fluctuate greatly; new variants often rapidly fixed
Typical morphological changes	Minor intensity; color, size, proportions, etc.	Similar to speciation but cumulatively greater in intensity; also polyisomerisms, anisomerisms, etc.	Pronounced or radical changes in mechanical and physiological systems
Typical population involved	Usually moderate with imperfectly isolated subdivisions	Typically large isolated units, with speciation proceeding simultaneously within units	Commonly small, wholly isolated units
Usual rate distribution	Erratic or comparable to horotelic rates	Bradytelic and horotelic	Tachytelic

A broad array of features underlie behavioral, physiological and reproductive adaptations. The conception of evolution is both humble and awe-inspiring; it is the sort of awe inspiration that Kant (1792/1951), the great 18th-century German Enlightenment philosopher/physicist, linked to his critique; a conception of awe in the face of nature, its diversity of expression.

BIOLOGY-AMAZING

Diverse information molecules have a long evolutionary history. Peptides and steroids and transmitters can be found in plants and insects. Plants, for example, have some of the essential information molecules. For instance, at night plants secrete melatonin, an information molecule that carries diverse forms of instructions and which is linked to circadian rhythms (Wehr *et al.*, 1993). Thus in plants, invertebrates and vertebrates, melatonin acts as part of the transduction mechanisms in adjusting to daylight and night time. In a diverse array of animals melatonin is linked to a number of physiological and behavioral adaptations.

Plants and fungi synthesize steroids that are known to regulate development and reproduction, respectively. Plant steroid hormones signal via membrane receptor kinases. The nuclear hormone receptors that underlie many of the actions of steroid hormones in mammals and other vertebrates are absent from plants, and are believed to have developed over 1 billion years ago after the divergence of the metazoans and fungi.

One fundamental form of nutritional and social attachment in mammals is lactation; the mammary glands, a cardinal feature of mammals, dates back to Triassic and Jurassic insectivores. In fact, spraying eggs with water may be an evolutionary precursor toward mammary gland development (Oftedal, 2002). And hormones such as angiotensin, vasopressin, oxytocin, prolactin, tied to fluid balance, are used in diverse functions, including the expansion of regulation that figures perhaps in the formation of new glands.

A kind of speciation occurs (retaining or not retaining a common function) by segregation (oxytocin, separated from vasopressin in different end organ systems), for instance, oxytocin within auditory regions of the brain at all levels of the neural axis (Kanwal *et al.*, 2002) could well be involved in animals that "see by hearing" in sensory processes utilizing vital ultrasonic vocalization coordination (Griffin, 1958). Evolution selects the diverse sensory systems that are linked to the organization of action. Oxytocin and vaso-pressin, expressed in auditory nuclei as in several species of bats in addition to a wide array of brain regions, could well be linked to be part of the systems in detecting auditory signals vital for approach (mother infant signals) or avoidance (predators [Carter *et al.*, 1997/1999]).

A number of mechanisms, e.g., peptides, include axon duplica-tion and loss, and exertion; gene duplication and variation on the pro-duction of these information molecules are traced back hundreds and millions of years and they then play diverse functions, e.g. food inges-tion and metabolism.

Conversely, anomalies are a wonderful reminder of humility. The platypus (Figure 1.2), an egg laying mammal, rich in informa-tion molecules (e.g. oxytocin, vasotocin, prolactin, angiotensin, etc.; Chauvet *et al.*, 1985) is an animal that speaks wonders about evolution or what Darwin (1859/1958) marveled at as "living fossils." Its diver-gence breaks a rule, in the same way orangutans are social isolates even though they are primates, and social life is the very definition of primates.

However, in the case of the platypus, we get a sense of our evo-lutionary past, the genomic makeup now understood and unique; it is phylogenetic history pregnant in the present (Warren *et al.*, 2008).

Mammals have mammary glands, rooted in lactation and tied in evolution. However, underlying lactation are the essential phylo-genetic ancient information molecules prolactin and oxytocin and angiotensin. Oxytocin is tied to successful hatching behavior, such as egg laying, and, as we will see in later chapters, social attachment. Prolactin also is knotted to social attachment (Wallis *et al.*, 2005). These information molecules (see Chapters 3–5) are at the heart and part of the arsenal that underlies the physiology of change and adaptation and of social attachment in social vertebrates, with both females and males participating in parental behaviors.

HOMINOID EVOLUTION

Primates are, for the most part, extensively social by necessity. Group formation is pivotal; since so much learning is required for our spe-cies, longer periods during ontogeny (Gould, 2002) are essential for learning the diverse local variation. A brain that takes time to develop and a nervous system with neurons that form diverse networks across

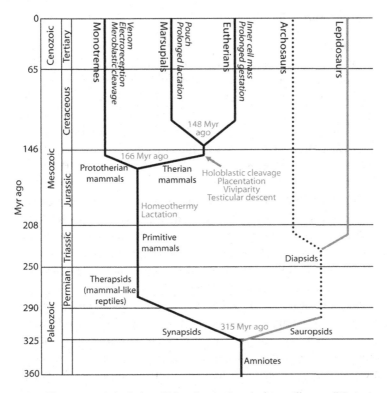

Figure 1.2 A depiction of the platypus' evolutionary lineage (Warren
et al., 2008) and the emergence of traits and mammals: amniotes split
into the sauropsids (leading to small birds and reptiles) and synapsids
(leading to mammal-like reptiles). These small early mammals developed
hair, homeothermy and lactation (black lines). Monotremes diverged
from the therian mammal lineage ~166 Myr ago and developed a
unique suite of characters (italic text). Therian mammals with common
characters split into marsupials and eutherians around 148 Myr ago
(italic text). Geological eras and periods with relative times (Myr ago)
are indicated on the left. Mammal lineages are in black; diapsid reptiles
shown as archosaurs (birds, crocodilians and dinosaurs) are dotted;
and lepidosaurs (snakes, lizards and relative) are in gray (Reprinted by
permission from Macmillan Publishers Ltd: Warren, W. C., Hillier, L. W.,
Marshall Graves, J. A., Birney, E., Ponting, C. P., Grützner, F., et al. [2008]).

the cephalic systems requires that cortical development in primates
expands in scope; the development of brain in ontogeny and prenatal
development reflects the degree of cognitive and social competence
that is necessary for successful survival.

Tools and time

Tool making must have taken a dramatic leap in human primate ancestors around 2.5 million years ago. This was no trivial event; fine motor control, and an expanded and extended use of the motor cortex and sub-motor cortical areas no doubt figured importantly in this evolutionary development. Bipedalism likely dates back to 4.5 million years or so and tool use about 2.5 million years on the fossil record (Foley, 2006). The search for a clear linear connection between *Homo habilis* through *H. ergaster* to *H. erectus*, for instance, through others and eventually to us, is still speculative.

However, that connection is not just physiological, but comes from a long history of tool making, standing erect, creating habitats, displaying diverse cognitive adaptations essential for enhanced learning of the social and ecological milieu. Figure 1.3 depicts an evolutionary trend from 4 million to 50,000 years with divergence of species and onset of tool use (Figure 1.3, Foley, 2006).

So core features in the origins of the genus *Homo* consist of some of the following: longer gestational period, long life spans, forward locomotion and heel and hind limb dominance during locomotion, dominance of stereoscopic vision and forward movement of the eyes, and expanding use of hands. The search for the origins of the genus *Homo* stretches back to the 1860s discoveries of Cro Magnon and Neanderthal remains, followed by Heidelberg man, and *H. erectus*, *habilis*, *ergaster* and *rudolfensis*. Depicted in the Table 1.2 are morphological features that are unique to *Homo* compared to other hominoids.

Many differences are obvious at a molar level across species; and what is different at a molar level of analysis, for instance, on the morphological side of *Homo* are changes in cortical expression, which

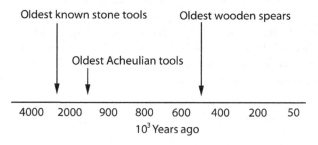

Figure 1.3   Some important temporal events for suggested tool use (adapted from Mellars 2006 and Zimmer 2005).

Table 1.2. *Defining characteristics of Homo (adapted from Wood, 1992)*

Characteristics unique to all species of Homo, *as distinct from other hominids:*
1. *Increased cranial vault thickness*
2. *Reduced postorbital constriction*
3. *Increased contribution of occipital bone to cranial sagittal arc length*
4. *Increased cranial vault height*
5. *More anterior foramen magnum*
6. *Reduced lower facial prognathism*
7. *Narrower tooth crowns, especially mandibular premolars*
8. *Reduction in length of the molar tooth row*

Characteristics shared by H. habilis *and* H. rudolfensis, *and unique to these species:*
1. *Elongated anterior basicranium*
2. *Higher cranial vault*
3. *Mesiodistally elongated first and second molars*
4. *Narrow mandibular fossa*

---

unlike changes in facial and dental morphology leave no fossilized traces (Wood, 1992).

What we do know something about, however, is the difference in cortical mass in diverse hominoids. The trend, not surprisingly, is towards greater weight across the evolution of *Homo*. Depicted in Table 1.3 are suggestions about cranial capacity, encephalization and brain size in extinct hominoids and humans (Foley, 2006).

The encephalization quotient (EQ) expresses brain size relative to body mass, and thus is a measure of relative brain size. Table 1.3 gives the EQ of various extant and fossil species.

One view of our evolution as a species is depicted in Table 1.4.

One is always struck by both the similarity and the differences of mammals (and vertebrates in general); the diverse forms of expression and the similarity of end organ systems is overwhelming evidence for the validity of evolution. Again, one system, however, always stands out – namely, the evolution of the central nervous system. At a molecular level of analysis the link between our species and other primates is amazingly profound (Evans *et al.*, 2006). But the morphological side, and diverse other features (rates of development, brain size, cognitive capacity) is amazingly different. A remarkable set of stable genes across species is commonplace; differences between our species and most other primates is quite small at the genetic level, and across mammals. Complete discernment of sequences reveals broad commonalities (Evans *et al.*, 2006).

Table 1.3. *The encephalization quotient (EQ) of various extant and fossil species (adapted from Power and Schulkin, 2009)*

Species	Date range	EQ
Australopithecus afarensis	3.9 – 3.0 Mya	2.5
Paranthropus boisei	2.3 – 1.4 Mya	2.7
Paranthropus robustus	1.9 – 1.4 Mya	3.0
Homo habilis	1.9 – 1.6 Mya	3.6
Homo ergaster	1.9 – 1.7 Mya	3.3
Homo erectus	1.8 Mya – 200 Kya	3.61
Homo heidelbergensis	700 – 250 Kya	5.26
Homo neanderthalensis	250 – 30 Kya	5.5
Homo sapiens	100 Kya – present	5.8

Table 1.4. *Approximate timeline for the succession of hominids (adapted from Donald, 1991)*

Time before present	Characteristics
5 million years: hominid line and chimpanzee split from common ancestor	
4 million years: oldest known australopithecine	• Erect Posture • Shared food • Division of labor • Nuclear family structure • Large number of children • Longer weaning period
2 million years: oldest known habilines	• As above, with crude cutting tools • Variable but larger brain size
1.5 million years: *homo erectus*	• Much larger brain • More elaborate tools • Migration out of Africa • Seasonal base camps • Use of fire, shelters
0.3 million years: archaic sapient humans	• Second major increase in brain size
0.05 million years: modern humans	• Anatomy of vocal tract begins to assume modern form

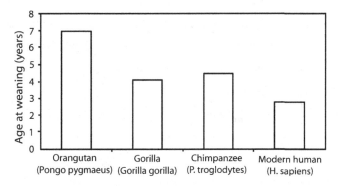

Figure 1.4   Weaning periods of us and related primates (adapted from Robson and Wood, 2008).

Most primates are highly social except for the solitary orangutan; of course there is a long relationship between the mother and her offspring and it is interesting to note the longer contact of mother and infant through weaning (Robson and Wood, 2008, Figure 1.4).

The exploration of diverse terrains, language and other cognitive communicative competence, omnivorous feeding and predatory patterns, tool use, and bipedal structure along with the massive social abilities figured importantly in our evolution and the colonization of the globe from our origins in Africa. These are core features of our evolution (Mellars, 2006).

Out of Africa

Figure 1.5 illustrates versions of modern depictions of dispersal out of Africa (Figure 1.5, Mellars, 2006).

Some of the key events and core features of our evolution over the last 300, 000 years are depicted in terms of key discoveries of modern human evolution in Table 1.5 (Zimmer, 2005).

Amid some presumably unusual and unstable environments, a cognitive expansion, a change in brain development, led to tool use and social cooperative behaviors that proved fateful for our development.

Some cognitive culture features in diverse homo are related to a common modern ancestor; social features essential for group formation are pervasive, as seen in Table 1.6.

Again, we know now that diverse forms of hominids competed and perhaps interbred during the same time period; *Homo sapiens* came

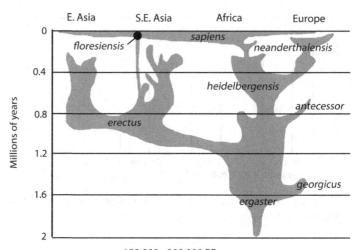

150,000 - 200,000 BP
Initial emergence of anatomically and genetically modern populations in Africa

110,000 - 90,000 BP
Temporary dispersal of anatomically modern populations (with Middle Paleolithic technology) from Africa to southwest Asia, associated with clear symbolic expression

80,000 - 70,000 BP
Rapid climatic and environmental changes in Africa

80,000 - 70,000 BP
Major technological, economic, and social changes in south and east Africa

70,000 - 60,000 BP
Major population expansion in Africa from small source area

ca. 60,000 BP
Dispersal of modern populations from Africa to Eurasia

Figure 1.5   Dispersion out of Africa (Mellars, P. (2006), going east: New genetic and archaeological perspectives on the modern human colonization of Eurasia. *Science, 313*, 796–800. Copyright (2006) National Academy of Sciences, U.S.A.) and some important events for modern humans during this same periods (Reprinted by permission from Macmillan Publishers Ltd: Foley, R., & Lahr, M. M. Human evolution writ small. *Nature, 431*, 1043–44, copyright (2004).

to dominate the landscape, as other human-like primates became extinct. In other words, the evolutionary perspective suggests that a wide range of human-like primates were existent at the same time (Foley, 2006). Evidence suggests that 30,000 years ago, Neanderthals and *H. sapiens* co-inhabited different but overlapping geographical

Table 1.5. *Another variant of evolutionary change (adapted from Mellars, 2006, and Zimmer, 2005)*

---

280 thousand years ago: Stone blades and ocher pigments used by ancestors of *Homo sapiens* in Africa.

195 thousand years ago: Oldest fossils of *H. sapiens* found in Ethiopia

140 thousand years ago: Evidence of long-distance exchange and shell fishing in Africa

130 thousand years ago: *H. sapiens* moves into the Levant

80 thousand years ago: Homo sapiens occupation of the Levant ends, replaced by Neanderthals

77 thousand years ago: Ornamental shells and geometrical engravings made in South Africa

50 thousand years ago: *H. sapiens* expands out of Africa

41 thousand years ago: Oldest evidence of humans in Europe

36 thousand years ago: Oldest Europeans cave art

33 thousand years ago: Neanderthal-made-ornaments suggest Neanderthals imitated *H. sapiens*

28 thousand years ago: Neanderthals become extinct

27–53 thousand years ago: Estimated age of the youngest *H. erectus* fossils in Indonesia

20 thousand years ago: *H. sapiens* moves from Asia into North America

18 thousand years ago: Youngest known fossils of *H. floresiensis*

---

locations in which climate change was a significant factor (Mellars, 2006).

Only one survived – us, Homo sapiens. But Neanderthals' brain morphology and social groups suggest high cognitive functions, and diverse forms of representative abilities. Mithen (2006) argues that the Neanderthals' symbolic burial sites and large brains suggest diverse cognitive abilities, including perhaps an elaborate form of song production without the expansion of linguistic competence that Homo sapiens expressed. Both forms of hominid shared a rich molecular biology and cognitive and social capacity.

Tool use in early hominids was pervasive; but then tool use is an ancient adaptation though the tools were crude. While the stick, compared to the telescope or microscope, might be crude, both expand animal horizons, and adaptation. Tool use requires an evolving motor system that runs the gamut of the brain and enhances the cognitive/ motor abilities that underlie the use of tools as extensions of bodily

Table 1.6. Hypothesized "cultural" properties of hominin taxa and some empirical bases (adapted from Foley, 2006)

	Common ancestor and australopithecines	H. ergaster	H. heidelbergensis	H. helmei / H. neanderthalensis	H. sapiens
"Cultural" inference		Greater planning depth Imitation Limited innovation Emotional affiliation? Theory of mind?	Parental care?	Greater planning depth Ethnic affiliation Symbolic thought Language	Strong ethnicity Extensive symbolism
Observation	Communities Fission-fusion Male kin bonding Territoriality Political alliances Basic tool making Ephemeral traditions	Significant meat eating Delayed growth Complex tool making Persistent traditions	Fire? Modern Life history?	Flexible technology Regionalization	Local networks Cultural replacements Rapid change Material diffusion

adaptations. What emerged in us, ultimately, were improved methods of passing information to others, including other generations.

Changing perspectives on progressive evolution: punctuated equilibrium

The classical view of evolutionary change is an upward movement of progression. This may not always be the case; we know now that evolution can be discontinuous. There can be rapid changes, and changes that are abrupt and discontinuous with what came before. Most evolutionary conceptions, including Darwin's, were unidirectional.

Punctuated equilibrium, in contrast, is an evolutionary thesis in which rapid change can be followed by long bouts of stasis. It also holds that evolutionary change is certainly not unidirectional. Some other evolutionary biologists before Eldridge and Gould suggested that evolution is not a steady stream of progression; Goldschmidt had suggested that evolutionary changes might be abrupt, but the prevailing view has been that speciation is a gradual and continuous process, as opposed to an abrupt change. Of course, the issue was always a bit more complicated with some features being understood as continuous and others not. There is still debate about the conception of punctuated equilibrium with regard to evolutionary change. It does appear, however, that evolutionary change is not always a continuous variable of progressive functions and expression; there are bouts of changes followed by stability. There, in fact, can be features of both punctuated change and bouts in progressive change (Foley, 2006). Interestingly, punctuated equilibrium, like the concept of allostasis, is about change. Both are regulatory concepts; one is about change at the species level, the other is about change at the level of whole body physiological level of analysis. In both cases, adaptation and viability are the short- and long-term considerations. Punctuated equilibrium suggests long bouts of stasis, maintaining equilibrium amid perhaps something like allostatic adaptation to change, in sudden unexpected contexts (Eldridge, 1985). Interestingly, "punctuated equilibrium" is a model of biological change in which bouts of staying the same predominates amid abrupt moments of change, and then maintenance of stasis. Allostasis is simply about adaptation to changing circumstances. But as Gould (2002, p. 767) notes, "stasis does not mean rock stability," and allostasis does not imply constant change.

CONCLUSION

All explanations of human behavior should, to some extent, be linked to evolution. The interesting evolutionary suggestion that has emerged is that several kinds of *homo* existed and competed at the same time; we survived, at least for now. But the march towards the expansion of brain capacity, a redistribution of regulatory function towards cranial dependency, had emerged. These events are knotted to our cognitive adaptations and our social and tool using capabilities; our ability to pool resources towards common anticipatory ends. Cephalic machinations are essential.

Evolutionary considerations are both exciting and humbling; there are always gaps in the fossil literature, never like the model of classical physics, and always vulnerable to misguided disputation. Both rapid and gradual changes are not inconsistent principles, and adapting to both change and coping with recurring features are also not mutually exclusive. Long-term viability is the measure of adaptation. Concepts such as allostasis emphasize the regulatory competence in adapting to changing social and ecological contexts, and not merely reacting and staying the same (Sterling and Eyer, 1998). Change and stability or stasis underlies evolution, in addition to speciation, variation, selection of behavioral and physiological characteristics, etc. However, the impact of a biological perspective can only be integral to a consideration of ourselves, our social groups formation and our epistemological orientation – a move away from absolutes towards humility and plausibility with an evolutionary consideration setting a context for understanding (Dewey, 1925/1989).

2

Cognitive Competence and Cortical Evolution

INTRODUCTION

Evolutionary factors leave their imprint across all life processes; one core event is the organization of the brain. Brains vary with species, with taxa; linking a conception of evolution with brain function required a conception of the nervous system, its core outline, variation and common themes (Swanson, 2000).

The nineteenth-century British neurologist J. Hughlings Jackson formulated a conception of the nervous system that places the brain within evolution; the brain is understood in terms of levels of function. The neocortex represented a crowning achievement, increasing the range of action.

This was, of course, understood for some time; Jackson's conception of the nervous system in the nineteenth century had cortical tissue, the evolutionary mantle of the brain, underlying the selection of options, in which brainstem sites carried out the basic functions. Devolution of function, typically for Jackson, was linked to forebrain damage, thereby limiting capacity and reducing behavioral options; what evolved was the range of opportunities that reflects the corticalization of function.

Of course, the mantle of progress with corticalization of function was at the center of an egocentric human conception of evolution, with social bonds, social history, human invention, rational decision forming being the pillars of social ascent. The march of progress was the conception, both in terms of biological and cultural evolution.

Two concepts are important for understanding our evolution with regard to the brain: one is the concept of corticalization of function and the other is the concept of the limbic brain. Corticalization of function is tied to our evolutionary ascent. As John Hughlings Jackson (1884/1956: p. 399) put it, our "highest cerebral centers ... evolved out of lower centers." In evolution, functions were represented at different levels of the neural axis (Gallistel, 1980).

Thus, an expanding sense of cortex has been readily understood for several centuries in the evolution of species; in the twentieth century, much detail was provided across diverse species with an expanding sense of detail and function (Rakic, 2002).

This chapter provides an orientation from which to engage the fundamental link between action, evolution and the social milieu. An expansion of cortical function and cognitive physiological functions, oriented to a complex social milieu: is an apparent feature of our evolution. This reflects our ability to react to stimuli and anticipate events and to regulate the internal milieu amidst a changing social context in an anticipatory (allostatic) manner (Sterling and Eyer, 1988).

CEPHALIC CORTICAL EXPANSION AND HOMINOID EVOLUTION

Diverse cognitive adaptations, including our ability to predict the behaviors of others, are a function of the fact that we tag our fellow humans in terms of their beliefs and desires. This, of course, is a higher order cognitive function. And we use that adaptation to, in part, predict what other human beings do in our social world, by their intentions.

A cognitive resource is this ability to track others by what we think they desire and believe. Of course, we track many behaviors that are simpler. For instance, what someone is looking at; joining eye contact on a common object, rooted together in a coordinated fashion, is at the heart of pedagogy. We learn from one another and manipulate one another and predict behaviors by what the focus is on, where the eyes are rotating towards, both externally and literally telling us something about beliefs and desires (Tomasello, 1999). Children are oriented within the first few months of life to form social contact through the visual system, to track events in a manner of joint attention to objects (Tomasello, 1999). These events are like social glue, facilitating future transactions with one another, and determining social adaptation. The social roots of our diverse cognitive capabilities are pervasive.

The demands of our long postnatal period are essential for pedagogy, during which sets of core cognitive capabilities are expressed. One core feature is the ability to determine whether an object is alive or not (Carey, 2009). Most, if not all, end organ systems have computational capabilities (e.g. kidney functions) in the maintenance of the internal milieu; but for our purposes in this context, we are talking about the integration of information from the external world, translated into coherent adaptive functions. Cognitive categories figure in our recognition of social and live objects. There is much unresolved debate with regard to the range, innateness and developmental expression of these capabilities. What is not debated is the fact that they are anchored to our social milieu: getting oriented to others, to the ecological and social surroundings. The range of computational and social skills that underlie our cognitive competence is nicely detailed in Table 2.1. Depicted are diverse tasks involving space perception, quantification, causation to social signaling, communication and theory of mind monitoring (Hermann et al., 2007).

The orientation of the child to a physical domain of objects can appear quite similar on some tasks to the common chimpanzee or orangutan in the first few years in development (Hermann et al., 2007); when given problems concerning objects in space, quantities or drawing inferences, chimpanzees and orangutans look similar to very young humans. What becomes quite evident early on in ontogeny is the link to the vastness of the social world in which the neonate is trying to secure a foothold (Figure 2.1).

Social intelligence, particularly in primates, is importantly bound to reproductive success; the alliances formed by mammalian females in a number of species, for instance, are vital for this (e.g. baboons, dolphins). A premium is set on cognitive evolution, an expression of diverse cognitive/behavioral adaptations coupled with cephalic expansion (Figure 2.2).

Diverse factors underlie the link between corticalization of function and both social and ecological factors in primate life, life span, group size, terrain adaptation, detection of predation, approach behaviors, foraging behaviors, etc.

Group size is linked to neocortex expansion in hominoids, as is longevity, as depicted in Figure 2.3. The pressure of coming into touch with others, creating alliances, and tracking lineages no doubt required more cortical mass (Dunbar and Schultz, 2007).

VISION, PRIMATES, JOINT SOCIAL CONTACT

Primates are such visual animals; we look at each other, watch closely what others do. In our species this trend contributed to our ability for

Table 2.1. *Diverse forms of cephalic capability (adapted from Hermann et al., 2007)*

Domain	Scale	Description
Physical	Space	Locating a reward
		Tracking of a reward after invisible displacement
		Tracking of a reward after a rotation manipulation
		Tracking of a reward after location changes
	Quantities	Discriminating quantity
		Discriminating quantity with added quantities
	Causality	Causal understanding of produced noise by hidden rewards
		Causal understanding of appearance change by hidden rewards
		Using a stick in order to retrieve a reward which is out of reach
		Understanding of functional and non-functional tool properties
Social	Social learning	Solving a simple but not obvious problem by observing a demonstrated solution
	Communication	Understanding communicative cues indicating a reward's hidden location
		Producing communicative gestures in order to retrieve a hidden reward
		Choosing communicative gestures considering the attentional state of the recipient
	Theory of mind	Following an actor's gaze direction to a target.
		Understanding what an actor intended to do (unsuccessfully)

joint contact, keeping tag of what we are both watching and of com-bining actions into coherent social organization.

Joint attention, or gaze following, is a fundamental adaptation demonstrated in a number of primates, and several studies have shown that diverse regions of the brain in primates, including neocortical and amygdala, are linked to gaze following. The temporal cortex and

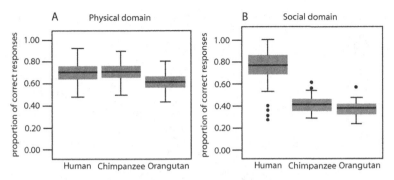

Figure 2.1 Young children and two related primates solutions to both simple physical and social problems. Physical domain (A) and social domain (B). The box plots show the full distribution of the proportion of correct responses for physical and social domains for each species (from Herrmann, E., Call, J., Hernadez-Lioreda, MV *et al.* (2007) Humans have evolved specialized skills of social cognition. Science **317**, 1360–6. Reprinted with permission from AAAS).

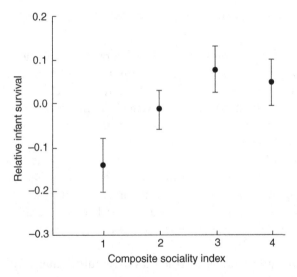

Figure 2.2   (From Silk, J. B. (2007). The adaptive value of sociality in mammalian groups. *Proc. Trans. R. Soc.*, **362**, 539–559. Reprinted with permission from AAAS.)

temporal pole and amygdala, long linked to meaning, are importantly involved in social contact and discernment of social meaning.

The visual cortex is, of course, essential for this, and projects to a region of the amygdala, linked to the formation of social

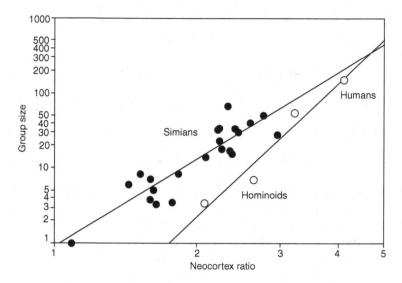

Figure 2.3   Group size in several primates and expansion of cortex
(reprinted from Dunbar, R. I. M. & Shultz, S. (2007). Evolution in the
social brain. *Science*, **317**, 1344–7 with permission from Elsevier).

attachment and social aversion are key neural connectivities. The
link to the amygdala (see Chapter 7) is through the lateral region,
and through to the central nucleus of the amygdala (LeDoux, 1995).
So, our cephalic systems are bulging with visual input, transformed
into vital pieces in the organization of action and the organization of
meaning (Barton, 2004).

A very early evolutionary trend towards stereoscopic binocu-
lar vision is linked to the expansion and design of the visual system
(Dunbar, 2003), and the evolution of the primate visual system brain.
The degree of corticalization of function is closely related to the anatomy of the visual systems; what stands out is the visual system in our
species.

An elaborate neural set of structures widely distributed throughout the brain is linked to keeping track of others, watching what they
do and gaining a foothold in the world of approachable and avoidable
events. An outstanding piece of our evolution is the volume of the
visual primary cortex as measured against the whole of the neocortex
(Dunbar and Shultz, 2007).

And, of course, brain size has long been linked to diverse cognitive functions, but is not simply a mammalian phenomenon; various
birds demonstrate remarkable cognitive abilities. One suggestion is

that the expansion of the basal ganglia is perhaps a variant of corticalization of function in the bird brain; telencephalic expansion has been explored as a phenomenon in the context of social complexity reaching back to dinosaurs. But for us, a wide variety of evidence links the degree of social interaction with neocortical expansion (Dunbar, 2003). Diverse models of group size have been linked to neocortical enlargement (Dunbar and Shultz, 2007) and cognitive competence (Byrne and Corp, 2004) across distributed cognitive systems (Barton, 2004).

COGNITIVE PREDILECTIONS

Social cooperative behaviors, in addition to an evolutionary arms race of cognitive capabilities, lie at our evolutionary base. Many core capabilities are depicted, such as numerical, spatial, theory of mind (predicting behaviors on the basis of their beliefs and desires), that are well-known ingredients of the human mind and to some extent other primates. But when the issue turns to social events, young children early on outdistance our closest primate relatives (Tomasello, 1999). The cognitive architecture is linked to making sense of our world. It is reflected in quite different kinds of events important to adaptation. Diverse cognitive systems are involved in the organization of action (Gallistel, 1980). Cognitive systems were, in part, selected to organize actions that underlie perception (Lakoff and Johnson, 1999); cognitive systems are not divorced from action but endemic to it (Table 2.2).

However, these ideas are not in a vacuum, they are in a context that is bodily in nature: struggling to learn something, persevering to acquire something, knowledge as a contact sport with others, getting linked to others, enjoying the solitude of one's self-enclosure amidst the safety of others, or despite others, and forming boundaries of protective parlance. Cognitive adaptation is in the doing of things for coherence of action in complex social environments, and for diverse cognitive adaptations to ecological conditions, and social communicative functions. We are rooted in diverse contexts about social objects, and diverse preadaptive cephalic systems are expanded in use in social contexts (Rozin, 1976).

Core cognitive architecture is mostly about kinds of objects (Lakoff and Johnson, 1999). While to be sure the linguistic competence of our species is an innate architecture with rich lexical hooks to semantic representations, language and the evolution of the neocortex is also linked to social behavior, social contact and social complexity (Dunbar, 1992). As social animals, we are oriented towards

Table 2.2. *Action and thought*

(Adapted from Lakoff and Johnson, 1999) Thinking is perceiving	(Adapted from Pinker, 1994). Ideas about:
Knowing is seeing	Living objects
Representing as doing	Objects and mechanics
Communicating as showing	Language
Searching as knowing	Probability
Imagining as moving	Food sources (avoidance and approach)
Attempting to gain knowledge is searching	Kinship relations
Becoming aware is noticing	
Impediments to knowledge are impediments to vision	
Knowing from a "perspective" is seeing from a point of view	

diverse expressions of our conspecifics that root us in the social world; for example, a world of acceptance and rejection, of approach and avoidance towards one another and towards social and ecological objects rich with significance and meaning (Cheney and Seyfarth, 1990, 2007).

The social world is full of signals of cognitive/behavioral significance that serve as an orientation in the organization of action. A broad-based set of findings in non-primates has been the link between social complexity and larger brain size (Byrne and Corp, 2004). The metabolic investment of larger brained animals is expensive; neural tissue is a high-energy organ. Brains expand while other tissue does not, or at least does not to the same degree. Interesting correlations have been suggested between neocortical size and social cognitive skills–Machiavellian skills. Recall Machiavelli, a sage of the Renaissance with a keen eye towards the human condition, education of leaders, and a common sense expectation of human decisions and base sensibility. Of course this is only one side of the cognitive social evolution. Detection and deception amidst cooperation and social prediction is a common occurrence that utilizes diverse cognitive systems (Byrne and Bates, 2007). Deception and neocortical expansion are depicted in Figure 2.4.

Our evolutionary ascent is bound to our social ability, in addition to tool making and the onset of linguistic competences. This is coupled with a long gestational period and the massive amount

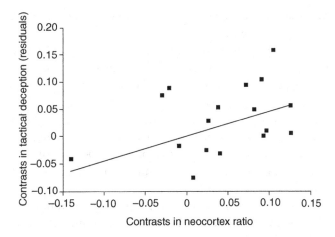

Figure 2.4   Deception and cortex: correlation between deception usage and neocortex ratio in primates. The frequency of within-group tactical deception was corrected for bias in observation effort, by using the residuals of the regression of deception against the number of studies (With permission, Byrne, R. W. & Corp, N. (2004). Neocortex size predicts deception rate in primates. *Proc. R.Soc.* **271**, 1693–9.)

of learning that takes place early in ontogeny with a long lactation period and long period of dependency. In addition, there is also a link between our longevity and the evolution of our problem-solving capabilities; our species had a greater opportunity to solve problems over time.

Moreover, brain capability is linked to an expansion of objects that are food sources, knowledge about where they are located, and other food-related behaviors, such as tending to and cooking it. Diverse foraging behavior and metabolism have been linked to an expansion of brain function (Aiello and Dunbar, 1993), and perhaps hominid evolutionary expansion. The brain is an expensive metabolic organ (of course they all are), and so a regulatory shift from the periphery to the central nervous system was an underlying adaptive factor for an expanding cortical mass (Power and Schulkin, 2009). An increased accessibility of brainstem function by cortical expansion is a basic blueprint of neural function in our species, and this cuts across a wide expansion of behavioral options. Moreover, food ingestion for us is highly social. Inhibitory roles of the cortex are pervasive, for example, functioning on basic appetitive mechanisms as we sit together, eat together, share our food, inhibit our responses, etc. Cortical enlargement can figure in many of the social contexts surrounding food: intake, cooperation, competition, prediction, etc.

Always the degree of cognitive competence and social gesture, bipedal organization, communicative engagement, diverse tool use and pedagogy are clearly linked to an expansion of the range of social contact (Dunbar, 2003). For example, the more grooming related behavior, reconciliation and social contact, the greater the degree of neocortical expansion, which may be particularly pronounced in females, in whom social contact is obviously linked to reproduction (Jolly, 1999). Of course, social reconciliation is a feature of many species, in both males and females (e.g. lions).

The important point in our evolution is the combination of not just deception, but trust and cooperation as important cognitive and regulatory adaptations (Jolly, 1999). Of course, trust and cooperation can enhance the use of deception. Competition is often overstated at the expense of cooperation; we readily cooperate to the benefit of our short- and long-term interests. With corticalization of function came an enhanced capacity to regulate the diverse competing social interests that interact with various motivational systems. But of course, the social milieu and adaptation are essential for successful reproduction, and as Darwin noted, "animals of many kinds are social" (Darwin, 1874, p. 96). But it is our species for which prosocial capabilities set the condition for diverse forms of moral sentiments.

REPRESENTATIONAL CAPACITIES FOR ACTION AND TOOLS

What emerged in our species was a highly linguistic, tool-using creature with social abilities and an elaborate diversity of cognitive skills. Many forms of instrumental expression and tool use arose, as did diverse forms of cognitive adaptations, some broad and some specific; cognitive expansion and fluidity from narrow adaptive abilities to expanding use across diverse problematic context is a feature of the human mind, e.g. speed versus diversity, flexibility and expansion of responses (Donald, 1991; Mithen, 1996). It is the expansion, as I indicated above, into more novel cognitive performance by increased accessibility to core cognitive functions by neural architecture that reflects our evolutionary ascent. Preadaptive cognitive functions are expanded in use, in diverse and broader contexts; the greater the connectivity, the less modular the access, the greater the expansion (Rozin, 1976). Key features of this cognitive fluidity are the integration

Table 2.3. *Core features of the mind (adapted from Mithen, 1996)*

• Social Intelligence
• Natural history intelligence
• Technical intelligence
• Language
• General intelligence

of several orientations to coping with the world: social intelligence, technical abilities, diverse expression of natural knowledge and of course language use, as seen in Table 2.3.

We are certainly linguistic animals. Language is a unique feature for our species with its rich syntactical elegance and expansion. Once syntactical language use emerged, our cognitive abilities increased in great measure, particularly our social discourse (Mellars, 2006). The premium of our evolution is on our social contact and discourse, our communicative competence and praxis. This social contact and preadaptation, in turn, is vital for the formation of basic regulative events that traverse a wide range of reward within the behavioral biology and the central nervous system (Carter *et al.*, 1997).

The representation of objects paralleled a desire to depict them in diverse ways, particularly those important to our everyday sustenance; depicting and predicting eatable objects is an important adaptation. A cognitive capacity to represent objects of diverse kinds is central to the evolution of the primate brain, particularly our species. Cognitive/behavioral capacities are, after all, ways to engage the world, to compute probability and assess friendly or non-friendly events. The representations are often social in nature, how to be linked with or to avoid others. We are interested in others, what they do; the representations are not simply divorced pictures from afar, they are active ways in which we engage the world. The outside world is important – the social world of others and our key to human flexibility, survival and reproductive success. Representation often has the connotation of something divorced from the object, something that detaches (though, of course, some forms of representations are indeed that way). But representations of others, of those that we care about, do not divide us from others; rather, they guide the organization of cognition, and the cognitive or informational systems are inherent to the organization of action (Gallistel, 1980).

This is important when we consider others, their beliefs and desires, the way they are oriented, to what they are oriented, learning from them, the tools they use, etc.

The traditional view made the cortex the only cognitive part of the brain (e.g. James, 1890/1952). Importantly, cognitive systems run across the brain; from cortex to brainstem, the central nervous system is knotted to cognitive rich information processing resources. One fundamental feature of our species is that our social discourse, social transactions, language, eating and tool use are embedded in the social milieu and not divorced from them; they evolved together in the more elaborate context of remaining in touch, of keeping track of one another. Language, checking in on others, gossip and social transactions were a recurrent and evolving set of features. Of course, the expansion of our motor capacity and our ability to see greatly co-evolved with these social capacities.

Interestingly, my old friend John Sabini, deceased now, a social psychologist and student of Stanley Milgram, often talked about the role of gossip for sharing social transgression, distasteful moral transgression and reinforcing social norms (Sabini and Silver, 1982; Dunbar, 1996). Gossip was a form of social glue, language related, socially oriented to tie individuals together. It is perhaps a linguistic expansion of grooming behaviors and larger social behaviors, tied to corticalization of function.

Now returning to our evolution, consider for a moment the time scale: we emerged about 100,000 years ago; agriculture developed in the range of 10,000 years ago; the move away from only being a hunter-gatherer and the domestication of dogs, and written language something in the order of 5,000 years ago; so amazingly close in time to where we are and so distant in what has occurred in this short interval. Cephalic accessibility is one evolutionary feature underlying this last fact with a common cutting theme between 30–60,000 years (Mellars, 2006).

Pregnant within the brain are diverse rich information processing systems that reflect an expanding cognitive capability. Diverse regions of the cortex are tied to motor function. The motor regions are rich with cognitive functions. Many species use diverse tools in adapting to their environment and reflect the evolution of cortical and subcortical systems in the brain that participate importantly in tool use, tool making and tool recognition. While I emphasize social behavior, tool construction and re-envisioning motor systems and the evolution of the brain are also significant; the diverse and expanding

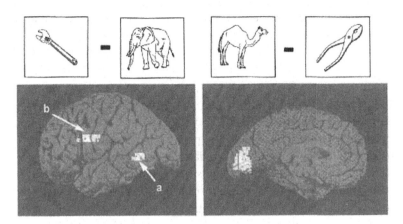

Figure 2.5   View of the left side of the brain showing areas in the
left posterior temporal lobe (A) and left premotor cortex (B) that were
more active when subjects silently named pictures of tools than when
they silently named pictures of animals. Reprinted by permission from
Macmillan Publishers Ltd: Martin, A., Wiggs, C. L., Ungerleider, L. G.,
& Haxby, J. V. (1996). Neural correlates of category specific knowledge.
Nature, **379**, 649–652, copyright (1996).

cognitive systems are present in motor regions of the brain (Barton,
2004) and are fundamental to our evolutionary ascent. Tool use is an
expression of an expanding cortical motor system in which cognitive systems are endemic to motor systems. Moreover, regions of the
brain are prepared to recognize differences between different kinds
of objects, one of which are mechanical tools. Importantly, frontal
motor regions have been linked to the motor features of tool use
(Martin *et al.*, 1996) (Figure 2.5).

Tool use and its elaboration is a reflection of expanding motor
capacity, rich with cognitive possibilities. Moreover we are not alone
in tool use; we watch birds use sticks to build nests, and chimpanzees and other primates use sticks to catch ants, or kill prey on occasion. The ability to use objects as tools facilitated the expansion of
our social milieu; the construction of tools is noted in diverse species,
but in primates, such as the chimpanzees, tool use is often tied to
a social context. And it is the expansion of cephalic function that
underlies the tool use that serves physiological/behavioral regulation.
An expanded motor system with diverse cognitive capacities no doubt
is pivotal in our evolutionary ascent. It is not just like the evolution
of the cortex or brain is knotted to social function; after all, tool use

and other diverse abilities are also knotted to the fact that we are social animals.

ALLOSTASIS, CORTEX AND SOCIAL FUNCTION

The concept of allostasis in its original meaning was tied to diverse and expanding social demands; social contexts that expanded and stretched mammalian capabilities, perhaps particularly in primates, beyond reflexive immediacy to deviations in set point physiology. In fact, a close look at regulation almost always reveals something more than simple reflexes to deviations from set point parameters. In the social domain, and with a consideration of evolution, enhanced use of behavior and social skills emerged to regulate viability. Corticalization of function is perhaps represented in the context of social intelligence and social regulation of the internal milieu; the utilization of others in the management of our needs, which have to do with things like safety and closeness, as well as the satisfaction of other basic needs such as human contact, are essential for survival and reproductive fitness.

In other words, our evolutionary ascent is inextricably linked to social complexity, to tagging the beliefs and desires of others as well as forging both cooperative and non-cooperative behaviors that reflect larger forms of regulatory competence and the ability for diverse patience. The greater the range of cooperative social behavior, the greater the modulation of the internal milieu by social contact, in addition to the greater ability to have diverse patience towards future preferences (Rosati et al., 2007). No doubt, one feature of this ability to restrain behavioral responses is in corticalization, a core idea since the nineteenth century. Within this shift to cortical capacity is restraining behavioral expression.

Allostatic regulation is tied to the physiology of change and adaptation, a core feature of our evolution and the structural formation of our cephalic expansion. The degree of corticalization of function corresponds to the expanded connectivity of cortex and diverse brain regions that underlie behavior. A greater form of regulation is adapting to changing circumstance, one mechanism of which is allostatic regulation. The expansion of anticipatory regulation of the social and the internal milieu is a reflection of allostasis. The move from reflexive to more anticipatory functions bound to a cognitive repertoire is a core feature of allostatic regulation of the internal milieu (Sterling and Eyer, 1988) (Figure 2.6).

A. Neocortical cascades to prefrontal cortex

B. Limbic cascades to prefrontal cortex

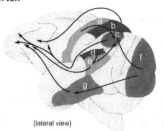

(lateral view)

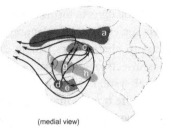

(medial view)

C. Prefrontal cascades to neocortex

D. Prefrontal cascades to limbic system

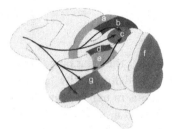

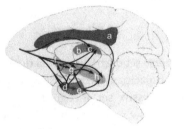

a. primary somatosensory
b. secondary somatosensory
c. inferior parietal lobe (multimodal)
d. primary auditory
e. secondary auditory
f. primary visual
g. secondary visual

a. cingulate gyrus
b. anterior thalamic nucleus
c. dorsomedial thalamic nucleus
d. amygdala
e. hippocampus
f. septum
g. hypothalamus
h. midbrain limbic area

Figure 2.6   Prefrontal cortex integrates cascaded inputs from neocortical and limbic systems, and feed back to both. This arrangement serves two functions: to imbue intellectual and cognitive context. Diagram shows the brain of macaque monkey (Sterling, 2004).

CAUTION AND HUMILITY: INVERTEBRATE AND VERTEBRATE BRAINS

The world is perceived depending upon the brain with which one is endowed; the fly brain, so radically different from our own, is still composed of diverse regulatory cephalic systems that underlie two fundamental events: feeding and reproduction (Dethier and Stellar, 1961/1970). The insect brain has no cortex and has a massively different set of features for so competent and elegant a species.

What was selected in these brains was speed and a limited, but very able, set of behavioral options. But diverse forms of learning can nevertheless be revealed (e.g. food source location), as well as a rich and diverse assortment of information molecules (e.g. diverse amines, including serotonin), which underlie development, metamorphoses that underlie diverse forms of social interaction, and division of labor in insect societies and fundamental changes in response to others (e.g. response to pheromones Dingle, 2002). In other words, invertebrates such as insects or even plants are also richly endowed with diverse hormonal systems that facilitate diverse events, including molting, and other life transitional stages and even parental behaviors (Dingle, 2002). In the next chapters, I will put some of these information molecules in perspective in terms of their functional role in vertebrate brains and for our evolutionary social trends.

CONCLUSION

A diverse set of cognitive functions underlies the great range of behavioral capacities that we display. There is no definitive list, but there is an impressive set of cognitive predilections in our species; the question is how many and how flexible they are in use and extension in novel domains. Over this there is much dispute. On the biological side, the cortical function underlies the great array of social and communicative cognitive-behavioral systems. In other words, cognitive capacity is linked with primate evolution to neocortical expansion; the flexibility that marks our species' behavioral repertoire options is a feature of corticalization of function (James, 1890/1952).

This cognitive competence, so social in its roots, also reflects the rapid brain growth during critical periods in our evolution, which perhaps figured in the dramatic expression of our social intelligence (Dunbar, 2003). Perhaps this resulted in a brain oriented to change, rather than stability. As such, our physiology may have changed to complement this cognitive change: namely from homeostasis, which means to stay the same, to allostasis, which is about the physiology and anticipation of change (Sterling and Eyer, 1988). The innervation of cephalic information processing systems into diverse peripheral sites underlies behavioral and physiological adaptation.

Anticipation of events, the capacity for foresight and traveling across time and important cognitive adaptations in our species' ascent, would be important in cooperative niches in which social milieu and its regulation figure importantly. Of course, in our species, planning for the future is a primary cognitive adaptation. Mental time frames are primary in planning ahead. In mammals, there is corticalization of function that underlies the expansion of cognitive adaptations;

perhaps regions such as the striatum maybe the analog in bird evolution to neocortical evolution in mammals.

Diverse anticipatory systems co-evolved in sustaining behavioral and physiological viability and are a feature of encephalization of brain function. And in the search for the "stable amidst the precarious" (Dewey, 1925/1989), human physiological regulatory systems depend upon both genetic and epigenetic forms of regulation (see Chapter 5). The integration of a genetic with an evolutionary perspective was a profound advance in our biological understanding (Mayr, 1963). Genetic programming amidst environmental conditions and possibilities sets the conditions for the possibility of adaptation.

Positive social contact in mammals, such as in our species, is as fundamental a feature as breathing oxygen; grooming behaviors are one kind of example in diverse mammals. Keeping track of others, forming alliances, and keeping track of the social milieu is the lifeblood of social commerce for which cognitive resources were expanded. These experiments linking long-term events and early social comfort provide a biological engine for understanding the events at a physiological/neural level of analysis. Both physiological and cognitive adaptation underlie the evolution of social knowledge and the regulation of internal milieu; knowing others, getting a link to the social world through them with fluctuating cooperation and social disappointment, is knotted to regulatory physiology and short- and longer-term viability (in other words, the circumstance of change) (Sterling and Eyer, 1988). Part of this exists in the prediction of con-specifics, something we as a species learn early and are fairly good at. Social groups are only partially about staying the same, but are about adapting to circumstances and anticipating social context as well.

3

A Window into the Brain

INTRODUCTION

As Darwin (1859/1958) noted, social instincts are the prelude for much of what governs our social evolution; climatic variation and stability perhaps fostered group formation as an important evolutionary adaptation – linking up with others, helping them, being helped. Of course the debate over the extent to which our behavior is governed by instincts has had a long and torturous history (James, 1890/1952). The social instincts are used to get linked to others and are bound to the development of a variety of social cognitive skills used to deceive (Mithen, 1996), trust, and engage in both social contact and social withdrawal. However, these instincts lie mostly in the formation of important alliances that underlie our cognitive evolution, amidst language and a set of diverse cognitive adaptations that root us in the world, a world in which epistemological and practical action predominate.

Both cultural and physiological/ecological stability are important factors in our evolution as are the diverse forms of adaptations that underlie coping with periods of non-stability. The search for stability and coherence amidst social insecurity is a fundamental feature that underlies human motivation and social investigation. In this chapter, I begin with an orientation of neuronal structure and function; a changing sense of our understanding of the organization of action through envisioning both the motor regions and limbic regions (e.g. information molecules that changed our conception of the classical limbic system) is vital for the expanding cognitive adaptations that figure in our social attachments and the allocation of cognitive resources for sustaining viable responses to changing landscapes. Yet the brain varies with evolution, and evolution is at the heart of understanding the brain, and our understanding of the brain has its own cultural evolution and history.

STRUCTURE OF BRAINS

Attempts to understand the structure and function of the brain date back thousands of years. Ancient and diverse cultures sought to depict the brain. The rise of modern science goes hand in hand with the evolution of the brain. Different kinds of animals were noted by anatomists to have had common themes, long noted before there was an evolutionary perspective of the brain in a biological context.

The anatomical inquisitiveness of the great Leonardo daVinci, and the elegance of Thomas Willis and many others, gave us morphological insight into the brain; a premium was, and is, placed in the depiction of the brain and other end organ systems with detail, accuracy and (on occasion) beauty of the structural depiction.

Seeds of an evolutionary understanding of the basic constituents of our planet are ancient and cyclical in expression. A conception of evolution is thus not new, but changed dramatically in the nineteenth century. Amidst a conception of evolution and the changing landscape of knowledge was a conception of brain design. Indeed, a fascination with brain design has long been expressed in human history. A comparative sensibility about brain size and cognitive capacity was adumbrated; depictions of the brain became commonplace.

The artistry of Santiago Ramón y Cajal (1906) as an anatomist parallels Da Vinci, but is quite striking in its own right; the artisan scientist infused the science of discovery with the culture of aesthetic sensibility. In this sense, the great Spaniard was the neuroscientist version of Da Vinci, who was also an anatomist. Amidst the discoveries about evolution of the brain, his elegant depiction and fusion of the artistic and the scientific would be a common theme for anatomists. Cajal's work led to modern notions of cortical design and neuronal organization (Rakic, 2002).

Cajal had a fundamental insight about layered neocortex and cortical columns, along with a more modern depiction; columnar organization became a foundational framework for neuroscientific research (Rakic, 2002). Moreover, Cajal laid out the concept of the neuron, the synapse, dendrite, glia, and electrophysiological mechanisms. The core morphology has been known for centuries in terms of spinal cord, cranial nerves, brainstem, cerebellum and forebrain; how to segment the structures of course continues in our semantic networks within neuroscience. But the diverse types of neural cells in terms of dendrites, synapses and core structure of cells have been constant.

At the level of the brainstem and the cranial nerves (all 12 of them) and the lower brainstem, structure is remarkably the same across mammals and indeed all vertebrates; the change is in the development of the forebrain.

The gustatory system is a vital part of the limbic brain, and begins at the level of the 7th, 9th, and 10th cranial nerves in the brainstem; there are 12 cranial nerves and the 10th nerve is the vagus (wandering) nerve that innervates most of the peripheral alimentary system. The alimentary system begins in the oral cavity and ends in the anus; the gustatory system is intimately involved in determining what is in the world and how things affect the gut (replenishment, e.g. Norgren, 1995).

The gustatory system makes its way to both regions of the brainstem (e.g. solitary nucleus, parabrachial region) and then off to forebrain sites: gustatory thalamus and gustatory neocortex for sensory related events and amygdala, hypothalamus and bed nucleus of the stria terminalis for viscerally related information tied to gustation (Norgren, 1995).

Sensory systems (or motor systems) reflect the diverse forms of adaptations; for us it is primarily vision, for dolphins it is audition, and for rodents it is smell. The thalamus and cerebellum have long shed their original conception as simply a way station to the cortex from the brainstem and the cerebellum simply motor region; these regions are richly computational at each level and region of the brain (e.g. learning). The septum, hippocampus, amygdala, and basal ganglia (nucleus accumbens within this region) are the cornerstones of the older conception of the limbic system and are old cortex (three layered as opposed to six) in some anatomical depictions; regions called the amygdala or basal ganglia are large heterogeneous regions (Swanson, 2000). These regions are tied to diverse functions that range from behavioral inhibition, to diverse forms of memory and attention (Berridge, 2007).

The evolved classical spinal cortical motor pathways, and the motor pathways in general, remain a wondrous thing of anatomical majesty. Long and direct, the organization of the motor and the visceral and sensory systems are direct and multi-synaptic and are at the heart of the organization of action and the regulation of bodily viability. The bulk of the basic plan has remained fairly intact in terms of our conception of it over the last 50 to 100 years across taxa for vertebrates in the background of an evolutionary perspective (James, 1890/1952). Nonetheless, diverse functions reflect the difficulty of processing sensory information from the brainstem where the cranial nerves send first order information through brainstem sites and then through the thalamus or other sites, some of which terminate in diverse regions of the cortex involved in sensory and motor functions and associative functions; it is just that all these areas are rich in information processing and, strictly speaking, carry cognitive capacity.

Importantly, many of these sites overlap with neuropeptides and neurotransmitters that underlie diverse organization for behavioral responses (Figure 3.1). These neuropeptides (e.g. vasopressin, oxytocin, endorphins) underlie the organization of the orientating responses to novelty, and to regulatory needs such as thirst and sodium appetite (Denton, 1982). Depicted below is a neuropeptide system (angiotensin) as it is expressed in the central nervous system and traverses brainstem sites as regions of the solitary nucleus, to forebrain regions such as regions of the amygdala, bed nucleus of the stria terminalis, hypothalamus, along with the central gustatory system (Davis, et al., 2010). Neuropeptides are part of the central processing rooted in our

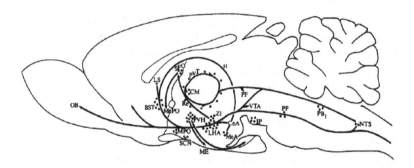

Figure 3.1   Schematic midsagittal view of rat brain illustrating the major angiotensin-II-immunoreactive cell groups and fiber pathways studied. Arrows are drawn to indicate the orientation of projection in cases where this is obvious. Note the NTS, nucleus of the solitary tract; PB parabrachial region; LHA, lateral hypothalamic region; SFO, subfornical area; BST, bed nucleus of the stria terminalis; CeA, MeA, central and medial amygdala (from Lind, R. W., Swanson, L. W., & Ganten, D. (1985). Organization of angiotensin II immunoreactive cells and fibers in the rat central nervous system. An immunohistochemical study. *Neuroendocrinology*, **40**, 2–24. Reproduced with permission of S. Karger AG, Basel).

orientation to basic functions, such as the motivation to ingest or avoid food resources.

Diverse regions of the brainstem are now understood as part of the limbic circuit (LeDoux, 1995). There are direct projections from the solitary nucleus to the amygdala (Norgren, 1995) where, for example, parts of the vagal nerve terminate, conveying site specific alimentary information. Moreover, information from the stomach, pancreas, duodena and liver is conveyed in site specific form to this brainstem region that projects to diverse regions of the brain including cortical and subcortical regions (Saper, 1995). Importantly, the brainstem massively projects to cortical regions, bypassing classical thalamic relay stations to cortical sites and then linking the peripheral systems with the central nervous system (Saper, 1995), in addition to the fact that all the information molecules are also produced in the brain (see Chapter 4).

Amazingly, the modern neural track tracing revolution that began in its modern form in the late 1970s has uncovered first and second order contact from cortical sites to these alimentary organs. The idea of a quick gut response has anatomical validity; the amygdala is connected to quick responses in getting the gist of things while invoking quick heuristic problem solving (Dewey, 1925; Adolphs, 1999). While the forebrain orchestrates the behavioral responses in a changing world of context and circumstance, the brainstem regions, such as the vagal complex, are vital for diverse forms of social attachment, in which neuropeptides such as oxytocin are vitally important (Carter et al., 1997). An orienting response is knotted to forebrain brainstem direct neuronal connectivity.

Arousal systems (formally linked to the reticular formation, Pfaff, 1999) underlie the orienting responses; these systems, as I indicated above, underlie a more general visceral system in the brain. Ascending activating systems have both general and specific functions. There are, of course, regulations of general arousal systems (e.g. adrenergic and serotoninergic systems in the brain) that underlie and sustain specific behavioral systems (sex behavior, hunger or thirst, social comfort, warmth, attention, learning). In other words, the generic features of the general arousal apparatus are broadly spread across diverse systems; more specific systems are much more targeted towards particular coordinated responses (Bodnar et al., 2002).

At each of these levels, there are important parts one could argue for the limbic system, and this is just one example: limbic areas are importantly involved in the regulation of approach and avoidance systems, but they are not confined to the classical circuit that includes regions of the amygdala and hypothalamus and part of the basal ganglia and cortex. The classical depiction of the arousal systems has been superseded by the anatomical depiction of transmitter and neuropeptide systems that underlie general arousal and specific functions; classical neurotransmitters such as norepinephrine, dopamine, serotonin and GABA underlie activation and quiescence. Neuropeptides traveling through many of the same regions are often colocalized with the neurotransmitters, which may underlie more specific behavioral functions. And thus, information molecules extend the concept of interconnectedness beyond local anatomy (Swanson, 2000).

It is the turning on and off of this central behavioral regulator that is essential for successful behavioral adaptation. The cortex is not simply about the inhibition of bestial expression, something often alluded to in the nineteenth centuries (James, 1890/1952), as "taming the beast." Cortical inhibition is often of less evolved areas (e.g. the amygdala), and of impulsive behaviors; hyperactivity of core areas has been linked to approach and avoidance behaviors. For example, fear and stimulating the amygdala (Rosen and Schulkin, 1998) underlie diverse forms of behaviors that are contextual.

An expanding cortex reflects an evolutionary adaptation in our species. But one should note an expanding piece of other tissue in, for instance, birds; social birds in particular have an evolving sense of subcortical tissue. Basal ganglia tissue may subserve some of the intellectual and social function that an expanding cortex does for mammalian tissue (Brown and Marsden, 1998). The frontal cortex is almost a third of our brain mass, but of course the visual cortex has expanded, as has all cortical tissue, except the olfactory cortex (Zald and Rauch, 2006). Re-envisioning the motor cortex and taking into account that the premotor cortex has imploded provides a refreshing new perspective of motor functions with no separation from diverse cognitive functions that stretch into basal ganglia function, long thought of as the head ganglia of diverse motor control.

In fact, Broca's area is tied to syntax, and regions of the basal ganglia underlie sentence construction. Wernicke's area underlies construction of meaning; patients with dopamine damage have deficits of the former to a greater degree than the latter (Ullman, 2004). Regions of the basal ganglia underlie the syntactical organization of behavior that Lashley (1951) theoretically posited and others have found (Berridge, 2004). Prefrontal and motor cortices take up a large amount of the cortex (Zald and Rauch, 2006). One core feature of the motor cortex is the organization of action, rich in cognitive resources

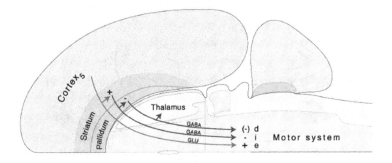

Figure 3.2   A modern depiction of corticostriatal brainstem projections (adapted from Swanson, 2000).

and connectivity to both cortical and subcortical regions of the brain (Swanson, 2000, Figure 3.2).

Some lexicons of what were the basal ganglia and largely subcortical are now understood as quite cortical (Swanson, 2000). Whether this anatomical conception stands over time is not certain; what is clear is those cognitive sorts of information processing, that are important for the organization of social action and social judgment, occur at the level of the basal ganglia. And many of these regions underlie social functions, including fast and rapid social assessment of events and mirroring actions, in addition to the lexicon and syntax of performative actions; fast motor action in learning to adjust is mirrored by neuronal activation in both frontal and basal ganglia function (Ullman, 2001). The broad repertoire of social perception and action is codified across diverse regions of the brain. A modern view of motor and cingulate cortex is depicted in Figure 3.3 (Rizzolatti and Luppino, 2001).

Action is rich in cognitive resources; the automatic perception of events, orchestration of action vital for diverse social behaviors has long been noted. One exception is that we have reenvisoned the motor regions below the neocortex with regard to the codification of action as we have expanded our notion of cognitive systems; there are many diverse cognitive systems that underlie the organization of action.

Some of these regions (Broca's and regions of the basal ganglia) are linked to the syntactical function of human language (Ullman, 2001). Basal ganglia function is linked to statistical computational competence, including events that are affect laden (Berridge, 2007). One recent experiment, for example, linked the valuation of others to social circumstance and comfort of living, and found greater activation

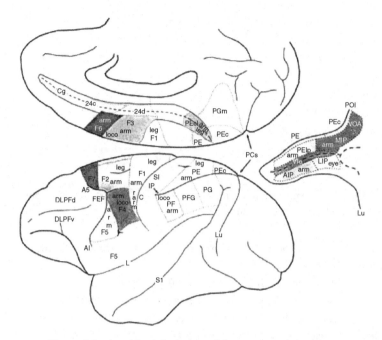

Figure 3.3   A modern depiction of the motor and cingulate cortex (Rizzolatti and Luppino, 2001); Medial and lateral views of the monkey brain showing of the motor cortex, posterior parietal, and cingulated cortices. The areas located within the intraparietal sulcus are shown in an unfolded view of the sulcus in the right part of the figure. A1 = inferior arcuate sulcus; AS = superior arcuate sulcus; C = central-sulcus; Cg = cingulated sulcus; DLPFd = dorsolateral prefrontal cortex, dorsal; P = principal sulcus; Pos = parieto-occipital sulcus; ST = superior temporal sulcus (from Neuron, **31**, G. Rizzolatti & G. Luppino, The cortical motor system, 889–901, copyright © 2001, with permission from Elsevier).

of the ventral striatum in an fMRI study (Fliessbach *et al.*, 2007); in diverse contexts of learning, tasks tied to reward have been connected to the human basal ganglia. In one such context, using fMRI, diverse regions of the striatum can be shown to be tied to learning of a task linked to reward from those who were deficient in the learning of a simple task (O'Doherty *et al.*, 2004).

From an adaptive social point of view, the expansion of basal ganglia function is linked to learning systems and prediction of events, with one set of dopamine-related neurons tied to uncertainty and the other tied to expectations of outcomes. In both cases, the anticipation of events is part of allostatic regulation, less on the reflexive response to change, and more on the anticipatory systems predicting changing events to remain viable (Schultz, 2007); this is essential in the formation of social groups, social cohesion and competition for diverse resources.

CHANGING CONCEPTIONS OF MOTOR REGIONS

Importantly, information processing systems permeate motor function and the basal ganglia, which is just one example amongst others. Re-envisioning this region of the brain is also seeing more links to traditional cortical tissue (Swanson, 2000). Massive connectivity to cortical and subcortical regions is a core feature of this traditional motor-related area, rich in many forms of cognitively fertile affective laden systems.

Re-envisioning motor/cognitive regions moves them across the full range of the neural axis with overlap on traditional regions underlying the emotions; regions of the basal ganglia are linked to regions that underlie approach and avoidance behaviors (Berridge, 2007). In one study, the anterior is more oriented to approach-related behaviors and the caudal end more oriented towards avoidance and negative responses. For instance, with injections of GABA in the rostral pole, a subset of the basal ganglia, the nucleus accumbens elicited feeding and approach related behaviors, as the injections of this neurotransmitter towards more rostral sites provoked behaviors turned towards rejection ingestive responses, avoidance and even fear related behaviors (Berridge, 2004).

RE-ENVISIONING THE LIMBIC SYSTEM

The idea of the limbic system expanded to a conception put forward by Paul MacLean (1990) who proposed a demarcation between neo-mammalian, paleomammalian and reptilian to account for the evolution of the brain.

This was an interesting perspective, which built on the work of many comparative anatomists who were behaviorally oriented, particularly C. J. Herrick (1926/1963). Each level of the neural axis represented something about the evolution of behavioral options important to adaptation; the crowning achievement is the expansion of the neo-mammalian cortex. For MacLean, the paleomammalian brain was tied to emotional adaptation, and limbic areas that underlie emotional responses to the world.

While we transcended the concept of the limbic system that was originally proposed (Broca, 1878), more and more areas of the brain became included, especially regions that underlie emotional adaptation (LeDoux, 1995). An evolutionary perspective for understanding behavioral functions is at the cornerstone of a neuroscientific perspective on social adaptation and well-being. Part of what broadened our conception of the limbic brain was the discovery of the chemical architecture of the central nervous system, namely that information molecules in the brain traversed the whole of the central nervous system, including those that underlie emotional or limbic adaptation.

The borders between what was limbic (and therefore tied to emotional adaptation) and what was not was much more expansive, not so expansive to perhaps eliminate the concept, but to see that it was a nominal distinction and not essential. Limbic areas run the gamut of the nervous system from first-order sensory information at the level of the brainstem and its projection to the forebrain; a good example is the gustatory system (Norgren, 1995).

ALLOSTATIC REGULATION AND CENTRIFUGAL CONTROL

The nervous system is an organ that drives order, and is tied to inherent and diverse oscillators in the brain. In turn, we are continuously adapting to the changing features of nature. Diverse forms of autonomic and sensory systems pregnant with cognitive resources underlie all forms of adaptive behaviors, including social behaviors rich in sensory information underlying contact and withdrawal (Stellar and Stellar, 1985).

The concept of centrifugal control is typically associated with top-down processing; stimulation of the amygdala expands the sensory fields of the cat to attack a rat or mouse. Different animals manage to coordinate action despite the influx of sensory information. Critical experiences are necessary for formation of the cephalic/behavioral engines that underlie the organization of action (Gallistel, 1980).

Each level of the nervous system is an information link to the organization of action, the formation of social bonds and aversions, innovation and discovery; cognitive systems do not begin with forebrain tissue. The functional architecture, not just the architecture frozen on the microscope, is embedded in action and adaptation in which the cognitive physiology orchestrates the behavioral adaptations; essential information molecules traversing the whole of the brain and utilizing common and divergent pathways are coordinated.

The anatomical revolution of the late twentieth century in part brought on the use of information molecules (e.g. angiotensin, oxytocin, CRH, see chapters 4–5) and the use of diverse viruses and other tract tracing materials to determine a chemical structure and distribution in the brain; the projections are massive from spinal cord brainstem to cortex. Massive projections cross the breadth and depth of the nervous system. As a result, classical pathways were redefined. Thus, the neural axis appears to be more distributed than strictly hierarchical in a Jacksonian manner (Gallistel, 1980; Berridge, 2007) that underlies the diverse social and ecological evaluations. Labile systems and distributed neural networks are a theme which we have redefined as strict hierarchical levels.

Levels of neural function underlie the organization of behavior; decerebralization leaves intact the basic oral facial reactions to gustatory information, such as basic licking or attraction to sucrose and rejection of quinine (Berridge, 2007). Jacksonian complete separation is a less reality as a clear demarcation of levels of function; nonetheless, the forebrain is essential in the initiation of diverse forms of motivational behaviors and particularly social behavior. The brainstem coordinates the basic reflexes and the forebrain orchestrates competing signals of importance in the organization of action (Gallistel, 1980).

These common architectural themes underlie most vertebrate architecture, with noted differences (Swanson, 2000). The cortex is expanded in mammals, the basal ganglia large in birds, and some marsupials do not have the massive connective tissue which we call the corpus collosum, which binds the two hemispheres in more than the anterior and posterior commissural connections. Consider further how our understanding of brain structure has evolved over time, and our understanding of brain function with regard to the social milieu and anticipatory regulation; our species devotes vast cognitive/neural capital towards the social milieu. Dolphins, another amazingly evolved species, known for their elaborate and long-term social bonds and complex social alliances, have an evolved auditory cortex and cerebellar cortices (Connor, 2007). They emit acoustic signals, varied and rich, and linked to forms of social cohesion. In other words, dolphins see the world through their auditory system, and this is reflected in their auditory cortex.

This in itself is not uncommon; less corticalized animals, e.g. the bat flying at night, emit sounds to find and avoid objects and see objects by hearing. Thus, complex alliances and a large social network are linked to corticalization of function. Of course, the social network of ants is a humbling reminder of how much social network one can achieve with no cortex and no limbic system.

If dolphins could communicate with us and had innate syntactical predilection about objects linked to audition, then perhaps they could do what a blind child appears to be able to do. It is an interesting fact that blind children still understand something about spatial objects and concepts. The sensory systems are pregnant with cognitive resources emboldening the organization of action vital for the formation of social contact and social aversion.

CONCLUSION

Though this is just a brief review of the topic, many core concepts and design principles of the brain have held fairly constant (concepts of synapse, axons, cell bodies, layered cortical design), while others have changed: the connectivity between brainstem and cortex, and the conception of function in the basal ganglia and cortex. Expanding concepts of the basal ganglia account for much more than the head ganglia of motor coordination; they are highly cognitive pieces of neural tissue involved in diverse capacities essential for the organization of action and for diverse social behaviors. What evolved in mammals is the cortex; what evolved in birds are the basal ganglia. But the basal ganglia are diverse, heterogeneous tissue.

We have changed our conception of the limbic system; it traverses the whole of the brain. Visceral projections rich in information reach telencephalic sites within one synapse. Getting the visceral sense of something, an object to approach or avoid is one synapse from the gut to cortex. Moreover, core information molecules such as peptides and neurotransmitters have diverse dispersion throughout the nervous system, often along the anatomical trajectory of major visceral and other sensory processes (e.g. gustation).

4

Chemical Messengers and the Physiology of Change and Adaptation

INTRODUCTION

Two fundamental ways of looking at the brain prevail: one is linked to chemical pathways of the brain, the other is electrical or biophysical law-like properties. One is not more valid than the other. They are just two rooted scientific approaches to understanding brain function.

Eliot Valenstein (2005), a behavioral neuroscientist at the University of Michigan, in his interesting book, *The War of the Soups and the Sparks*, lays out the discovery process involved in the depiction of the classical neurotransmitters (norepinephrine, dopamine) and the depiction of the sympathetic and parasympathetic systems and their representation in the brain. This insight into the chemical nature of the classical neurotransmitters along with the depiction of the electrophysiological properties set the conditions for understanding the brain.

My focus is on the chemical pathways of the brain and the functional role in the organization of social approach and avoidance behavioral responses (adaptive responses rooted in evolution). Social contact, in addition to the development of motor control, language and other cognitive capacities, are critical factors in our evolution. Cave paintings show a sense of the intellectual quest.

An understanding of the sympathetic and parasympathetic systems represented in both the peripheral and central nervous system underlie social attachments and all forms of regulation. A rich chemical milieu permeates these autonomic peripheral systems that underlie the diverse forms of physiological systems across end organ systems in maintaining long-term physiological stability.

Importantly, the chemical messages are richly expressed in the central nervous system.

Social contact again is a major adaptation in the regulation of the internal milieu, a way in which the information molecules that permeate cephalic and end organ bodily systems interact and result in behavioral and physiological activity; many of the same information molecules that are produced in the periphery (e.g. stomach, heart) are produced in the brain. There is a quite diverse set of them (e.g. CRH, oxytocin, vasopressin, angiotensin, prolactin, melatonin, serotonin, GABA, insulin, leptin, ghrelin). In this chapter, I continue the orientation towards the diverse forms of information molecules (e.g. angiotensin, corticosteroids, CRH), and their implication into understanding the organization of action and adaptation to the social milieu.

ORIGINS

Life on this planet began an enormously long time ago. Change seems slow, compared to 100,000 years ago, and the rise of this primate (H. sapiens, us). But it took millions and millions of years; the time scale is enormous. This new period is so short in time and so profound in human change, but change is the underlying theme in evolution along with adaptation; viability of outcome promotes long-term longevity.

One depiction of the different periods of time in which different vertebrates made their presence and began a process of evolutionary change is depicted in Table 4.1.

Species came and went; many went as we discussed in Chapter 1. Some are remarkably old (e.g. crocodiles) while some are relatively new (e.g. primates). One common thread among them is that they all possess diverse information molecules such as peptide hormones (Table 4.2).

In mammals, these information molecules became oxytocin and vasopressin; many of their functions are originally knotted to fluid balance, but they diversified in function. Many of these information molecules are quite ancient in function; consider gonadotrophin releasing hormone, which dates back to jawless fish, a common feature across the basic information molecules in the brain (Strand, 1999).

Evolutionary expansion is in part reflected in the polymorphisms of peptides and their use across diverse organ systems. The vasopressin family is quite ancient and is bound to vasotocin and isotocin/oxytocin (Strand, 1999), and indeed the hormonal cascades are quite old, and are a core feature.

Table 4.1. *Stages in evolution (adapted from Bentley, 1982)*

Era	Period	Millions of years ago
Cenozoic	Quaternary	1-recent
	Tertiary	70-1
Mesozoic		135-70
	Jurassic	180-135
	Triassic	225-180
Paleozoic	Permian	270-225
	Carboniferous	350-270
	Devonian	400-350
	Silurian	440-400
	Ordovician	500-440
	Cambrian	600-500

Table 4.2. *Some "vertebrate" neuropeptides found in unicellular organisms, plants, and invertebrates (Strand, 1999)*

Protozoa	Bacteria	Fungi	Plants	Invertebrates
ACTH	ACTH		TRH	ACTH/MSH
β-Endorphin	β-Endorphin		β-Endorphin	β-Endorphin
Insulin	Insulin	Insulin		Insulin
Gonadotropin	Gonadotropin			Glucagon
CCK				Gastrin
Calcitonin				GnRH
Somatostatin				Bombesin
				CCK
Glucagon				Angiotensin I
Vasotocin				ANH
				Pancreatic
				peptide
				Vasopressin
				VIP

    The genes that underlie the production of information molecules, such as peptides and steroids, stretch back millions of years (Strand, 1999). Molecular biology linked a number of steroid hormones, including aldosterone, corticosterone, and vitamin D, within a common molecular ancestor gene. Vitamin D is a steroid hormone expressed in diverse end organ systems including the brain. An evolutionary perspective is always critical in understanding their role in behavior and physiology.

There are diverse endocrine systems which secrete information molecules, some of which are listed in Table 4.3.

CHEMICAL MESSENGERS, INFORMATION MOLECULES
AND THE BRAIN

A key discovery in the brain is the expression of diverse chemical messengers that transmit information that is fundamental for both short- and long-term viability and or adaptation. There are diverse chemical signaling systems in the brain, some of which are neurotransmitters, including dopamine, norepinephrine and serotonin. These are broad-based chemical signals that regulate broad-based behavioral and physiological responses. Another neurotransmitter in the brain is GABA, an important inhibitory neuron colocalized with a number of other transmitters (Swanson, 2000). Other neurotransmitters are listed in Table 4.4.

Neurotransmitters tend to be broad in terms of regulation; for example, dopamine is importantly knotted to the organization of action, the prediction of reward, the mobilization for effort (Schultz, 2007). We usually associate dopamine or diminished regulation and expression with Parkinson's disease. Dopamine is of course produced not just in the brain but also in the adrenal gland, but the transmitter underlies all aspects of the organization of action. Hormonal messengers are a fundamental information class molecule of which there are many. Dopamine and norepinephrine or serotonin when expressed in peripheral tissue, such as the adrenal gland, are thought of as hormones; in the brain they are neurotransmitters. They are the same molecules involved in the organization of behavior as physiological signaling systems (Herbert and Schulkin, 2002).

All steroids are derived from cholesterol; they are mostly produced outside the central nervous system, though low levels of progesterone are produced in the brain (McEwen, 1995). Importantly steroids cross the blood–brain barrier (unlike peptides, which generally do not), and enter the brain through circumventricular organs (regions of the brain such as the subfornical organ). Common ancestor genes underlie diverse steroid hormones. Cholesterol metabolism is ancient.

Table 4.3. *Glands and hormones*

Endocrine gland	Major hormones secreted
Anterior pituitary	Growth hormone
	Prolactin
	Adrenocorticotropin (ACTH)
	Luteinizing hormone (LH)
	Thyroid-stimulating hormone (TSH)
	Follicle-stimulating hormone (FSH)
Neurointermediate lobe/ posterior pituitary	Arginine vasopressin
	Oxytocin
	Endorphins, enkephalins
Pineal	Melatonin
Thyroid gland	Thyroxine
	Calcitonin
Parathyroid gland	Parathyroid hormone
Heart	Atrial natriuretic factor
Adrenal cortex	Glucocorticoids
	Mineralocorticoids
	Androgens
Adrenal medulla	Epinephrine
	Norepinephrine
Kidney	Renin
Skin/kidney	Vitamin D
Liver/lung	Preangiotensin/angiotensin
Pancreas	Insulin
	Glucagon
Stomach and intestines	Cholecystokinin
	Vasoactive intestinal peptide
	Bombesin
	Somatostatin
Gonads: ovary	Estrogen
	Progesterone
Gonads: testis	Testosterone
Macrophage, lymphocytes	Cytokines
Adipose	leptin
	Ghrelin

Table 4.4. *Classic neurotransmitters in the CNS*

Catecholamines
Dopamine
Norepinephrine
Epinephrine
Indoles
Serotonin
Melatonin
Cholinergic
Acetylcholine
Amino acids
γ-aminobutyric acid (GABA)
Glutamate
Aspartate

Peptide hormones, as opposed to steroid hormones, are produced in the periphery and also produced in the central nervous system; in the periphery they are peptides, and in the central nervous system they are called neuropeptides. Hormones such as oxytocin, prolactin, vasopressin or angiotensin, are knotted to peripheral function tied to parturition, lactation and fluid volume, and are also expressed in extra hypothalamic/pituitary sites in the brain. They are derived from proteins, and have common and divergent evolutionary histories (Table 4.5).

The information molecules such as those listed above underlie a great deal of behavioral and physiological regulation; they are ancient and well expressed, and are the building blocks of diverse signaling features in both the central and peripheral nervous system as well as in end organ systems. For instance, neurotransmitters or neuropeptides convey visceral information to the brain across many areas (Sawchenko et al., 1984). CRH is a good example; diverse forms of visceral and inflammatory distress activate CRH gene expression across peripheral end organ systems in addition to central nervous system tissue (Yao et al., 2008). But nothing is quite like the paraventricular nucleus for the wide region of peptide production and regulation: the paraventricular region of the hypothalamus depicted by my friend and colleague David Jacobowitz (1988, Figure 4.1) at the National Institute of Mental Health. (Figure 4.1)

As I indicated in the previous chapter, many of these neural circuits have replaced what was once called the reticular activating

Table 4.5. *Some major neuropeptides and proteins*

β-endorphin
Dynorphin
Enkephalin
Somatostatin
Corticotrophin-releasing hormone
Urocortin
Atrial natriuretic factor
Bombesin
Glucagon
Vasoactive intestinal polypeptide
Vasotocin
Substance P
Neuropeptide Y
Neurotensin
Galanin
Calcitonin
Cholecystokinin
Oxytocin
Prolactin
Vasopressin
Angiotensin
Interleukin
Thyrotropin-releasing hormone
Gonadotropin-releasing hormone (GRH)
Luteinizing-hormone-releasing hormone (LHRH)
Neurotropin
Calretinin
Leptin

system, since many of these information molecules arouse and calm the brain. Cortisol, a wake up hormone, and prolactin are linked to many functions for species that rest. Melatonin also may be linked to quiescence (Wehr *et al.*, 1993).

Depending on where peptides are expressed, they have different functions; in extrahypothalamic sites in the brain, they are often linked to specific kinds of behaviors. Peptides are derived from proteins, and thus neuropeptides are bound to more specific functions (angiotensin and thirst) than neurotransmitters (dopamine and the organization of movement, serotonin and the tonic tone, which permeates central nervous system function); the classical hypothalamic

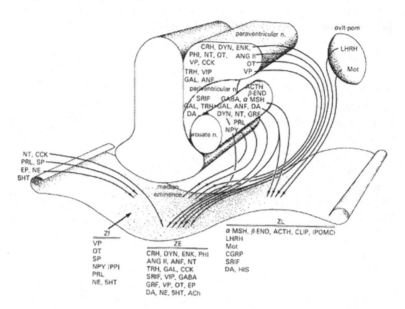

Figure 4.1   The many information molecules in the paraventricular nucleus of the hypothalamus nicely depicted by David Jacobowitz (1988) (Jacobowitz, D. M. (1988). Multifactorial control of pituitary hormone secretion: The "wheels" of the brain. *Synapse*, **2**, 86–92. Copyright (1988) Reprinted with permission of John Wiley & Sons).

pituitary adrenal axis is the normal semantic association, but they range throughout the central nervous system.

Look at the massive cortex; the frontal cortex is about one-third of the brain. Look at the hypothalamus and pituitary gland in comparison; the hypothalamic pituitary thyroid axis, or hypothalamic pituitary gonadal axis, of course look similar in size and negative feedback dominance (though not exclusively) in function.

EVOLUTIONARY EXPRESSION OF CORTICOSTEROIDS AND CRH

One example, dear to my experimental heart, are the adrenal steroid hormones; my first project was on the adrenal steroid aldosterone and its role in the behavioral regulation of sodium (Richter, 1943). Since that time, our understanding of the many diverse roles of aldosterone has expanded beyond the narrow, though life-sustaining role, that it plays in sodium and body fluid homeostasis (Schulkin, 2003). Later, it was discovered with regard to fluid balance that one cephalic mechanism was via the induction of angiotensin (Denton, 1982), and that

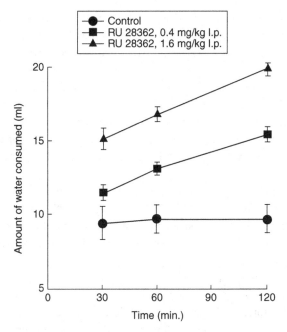

Figure 4.2   Water ingestion following systemic daily injections of the glucocorticoid agonist followed by a central injection of angiotensin II (reprinted from Sumners, C., Gault, T.R., and Fregly, M.J. (1991). Potentiation of angiotensin II-induced drinking by glucocorticoids is a specific glucocorticoid Type II receptor (GR)-mediated event. *Brain Res*, **552**, 2, 8. Copyright (1991), with permission from Elsevier).

both adrenal steroids (Wolf, 1964) facilitate angiotensin expression in the brain and facilitate the behavioral adaptation of water and sodium ingestion (Figure 4.2).

Corticosterone is the better known adrenal steroid and the central expression of CRH. This has been a big part of my research program for some time and is one example of steroid neuropeptide interactions in the organization of physiological and behavioral adaptations. It also is a window into how one information molecule, in this case CRH, serves diverse functions through a number of end organ systems. The ability to synthesize corticosteroids (e.g. aldosterone, corticosterone) probably arose prior to the origin of the jawed vertebrates. The ancestral glucocorticoids and its descendants were structurally preadapted to bind aldosterone, and thus 'exploited' the ligand that appeared around 200 million years after the receptor (Lovejoy and Jahan, 2004). This suggests that as few as two amino acid substitutions were necessary in the glucocorticoid lineage to generate a protein with high affinity for glucocorticoids but low affinity for aldosterone (De Kloet, 1991). The existence of two distinct hormone-receptor pairs allowed for the evolution of specific endocrine control of osmoregulation and adaptation to changing contexts. One allowed for an adaptation for the colonization of land (aldosterone). In many contexts, the two steroid potentiate the same behavioral and physiological events (e.g. fluid balance), while they compete for the same receptor sites in diverse sites in the body, including the brain (McEwen, 1995).

Tissue capable of corticosteroidogenesis is found in extant chondrichthyean fishes such as the holocephalans (ratfish) and elasmobranchs (sharks, skates, rays). In modern bony fishes, the teleosts, cortisol is the major steroid produced by adrenocortical tissue (Bentley, 1982) (Figure 4.3).

Corticotrophin-releasing factor (CRH) is the major neurohormone regulating the vertebrate endocrine stress response. CRH is a 41-amino acid peptide originally isolated from ovine hypothalamus and found to be the major hypothalamic releasing factor for pituitary corticotrophin (ACTH; Vale *et al.*, 1981) richly expressed across phyla (Table 4.6).

The classical hypothalamic pituitary adrenal axis in vertebrates is well-known; CRH secreted by the parvocellular acts on the pituitary to release ACTH, which then acts on the adrenal gland to release corticotrophin, which then circulates and returns to the brain to restrain or limit its own release (Dallman *et al.*, 1987). This is classical negative restraint, and it holds for most of the hypothalamic pituitary end organ systems. The HPA regulation is a fundamental way in which the brain organizes peripheral end organ systems tied to diverse systems (HPG, HPT, etc.).

These regulatory systems are phylogenetically ancient. The interregnal gland serves as an adrenal gland in the production of adrenal steroid hormone, both glucocorticoids and mineralocorticoids (Bentley, 1982); they then serve to regulate diverse end organ systems including CRF (or as it is called by many of us CRH in the brain). The adrenal gland not surprisingly, like most end organ systems, evolved in form and varied in structure as species colonized land.

Coinciding with the discovery of CRH, structurally similar peptides were isolated from the caudal neurosecretory organ (urophysis) of several species of fish (urotensin I) and from the skin of the frog *Phyllomedusa sauvageii* (sauvagine) (Denver, 2009). CRH and related

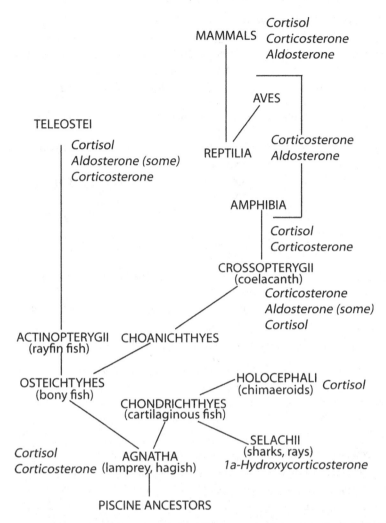

Figure 4.3   Aldosterone and corticosterone in diverse species (adapted from Bentley, 1982).

molecular sequences are ancient; they appear in invertebrates' and chordates' ancestral genes and therefore appeared very early in animal physiology, even before the appearance of jawed fish (Lovejoy and Jahan, 2006). Urocortin is a related peptide to CRH, and duplication of genes can appear even before there is a known receptor.

A series of studies in the brains of frogs demonstrates that the expression of these information molecules are ancient and play a role in the conversion processes that occur in metamorphosis and

Table 4.6. *Corticotropin-releasing hormone sequences*

Perciformes	Tilapia	SEDPPISLDLTFHLLREMMEMSRAEQLAQQAQNNR RMMELF
Salmoniformes	Rainbow trout	SDDPPISLDLTFHMLRQMMEMSRAEQLQQQAHSN RKMMEIF
Siluriformes	Bullhead	SEDPPISLDLTFHLLREMMEMSRAEQLAQQAQNNR RMMELF
Cypriniformes	Sucker	SEEPPISLDLTFHLLREVLEMARAEQLVQQAHSNR KMMEIF
	Carp	SEEAPISLDLTFHLLREVLEMARAEQMAQQAHSNR KMMEIF
	Goldfish	SEEPPISLDLTFHLLREVLEMARAEQMAQQAHSNR KMMEIF
Amphibia	Xenopus	AEEPPISLDLTFHLLREVLEMARAEQIAQQAHSNR KLMDII
Mammalia	Sheep	SQEPPISLDLTFHLLREVLEMTKADQLAQQAHSNR KLLDIA
	Cow	SQEPPISLDLTFHLLREVLEMTKADQLAQQAHNNR KLLDIA
	Human	SEEPPISLDLTFHLLREVLEMARAEQLAQQAHSNR KLMEII
	Dog	SEEPPISLDLTFHLLREVLEMPGAEQLAQQAHSNR KLMEII

are conserved across and expanded across species and phyla. CRH is also activated in diverse forms of external events in, for example, the frog *Xenopus laevis*, and the distribution of CRH and GR (glucocorticoid) immunoreactivity in the frog brain is evolutionarily conserved with mammals. Glucocorticoids are colocalized with CRH, and response components are in the proximal promoters of the frog CRH genes (Denver, 2009).

While glucocorticoids are well known to exert negative feedback on PVN CRH neurons in mammals (Sawchenko, 1987), there is similar regulation in amphibians. We found a similar relationship in the frog, where CRH was increased by treatment with CORT in the MeA and BNST. The positive regulation of CRH by glucocorticoids could be, at least in part, direct on CRH expressing neurons since CRH and GR were found to colocalize in these cells in the rat brain (Sawchenko, 1987). In the frog brain, CRH and GR are both expressed in the MeA and BNST in this amphibian as opposed to mammalian rat central nucleus of the amygdala and lateral bed nucleus of the stria terminalis (Denver, 2009).

As in mammals, glucocorticoids decreased CRH expression in the POA, which is the major site of hypophysiotropic CRH neurons in the frog and is homologous to the mammalian PVN. By contrast, glucocorticoids elevated CRH in the frog MeA and BNST (see also Chapters 5 and 6). CRH and its receptors are widely distributed in the brain and across different end organ systems outside the nervous system in a number of species (Yao *et al.*, 2008). Some clarity about the link across species between the CRH receptors has been outlined (Denver, 2009).

Importantly, in addition to the hypothalamus, CRH and related peptides are widely expressed throughout the mammalian CNS in limbic structures, the cortex, and brainstem nuclei associated with autonomic functions. In the CNS, CRH is proposed to function as a neurotransmitter, and while there are several types of CRH receptors, the TYPE 1 is tightly linked to diverse forms of behavioral expression and regulation (Denver, 2009).

Indeed, CRH is linked to diverse behaviors: food intake, learning, arousal, and sexual behavior. In the brainstem, CRH influences blood pressure and locomotor activity. Thus, CRH not only acts as a neurohormone in the HPA axis, but also integrates autonomic and behavioral responses to stress through its actions within the brain. And it is CRH often produced outside the PVN that seems essential for behavioral regulation (Davis *et al.*, 1997; Davis *et al.*, 2010).

However, many peptide hormones that were once thought to be restricted to the brain and pituitary are now known to be widely expressed in peripheral tissues, and CRH is just one of many. It is now known that the signaling components of the CRH system are widely expressed in peripheral tissues of mammals and frogs, suggesting that the expression and function of these peptides outside of the central nervous systems and pituitary gland arose early in vertebrate evolution. These peptides may influence most, if not all, physiological systems, including the nervous, endocrine, vascular, cardiovascular, skeletomuscular and reproductive systems. The heart, spleen, pancreas, kidney, liver, skin, GI tract, lung and gonads, among others, express CRH-like peptides, CRH receptors and CRH binding proteins (Denver, 2009).

For instance, CRH-like peptides are also located on the heart and the vasculature. CRH applied directly to isolated heart preparations increased coronary output and the release of atrial natriuretic peptides. Indeed, CRH peptides are known to be vasoactive and data show that they play important roles in cardiovascular regulation (Nemeroff *et al.*, 1992). Importantly, in many species (primates, rodents, reptiles, amphibians), CRH is expressed in the amygdala, BNST and medial pallium, and its expression, and that of the immediate early gene *c-fos*, are increased following exposure to a stressor (Denver, 2009). CRH is well known across phyla to promote locomotor activity (Moore and Rose, 2002). Diverse social and internal systems activate gene expression underlying CRH and many other information molecules in the brain. In other words, like other neuropeptides (e.g. angiotensin, enkephalin, neuropeptide Y, oxytocin, prolactin and leptin) there is a wide distribution throughout the brain and diverse end organ systems and a conserved expression.

FROM NEGATIVE RESTRAINT TO ALLOSTATIC REGULATION:
BEYOND THE HPA AXIS

The inhibition of CRH by cortisol is profound, though it is not always expressed under conditions of adversity. Indeed, many forms of adversity in which cortisol is elevated do not result in the inhibition of CRH in the parvocellular region (e.g. withdrawal from psychotropic drugs, restraint stress). Moreover, there are CRH neurons, particularly those that project to the locus coeruleus, which are unregulated by elevated infusions of cortisol (Watts and Sanchez-Watts, 1995). The mechanisms for this differential regulation still remain elusive, though regulation of transcription of factors located in the CRH promoter may include CREB, Fos, Hun and other intracellular factors (Yao *et al.*, 2008). In an important study by Shepard *et al.* (2005) adrenalectomized rats without elevated corticosterone, subjected to restraint stress, nonetheless showed the same pattern to CRH mRNA to restraint stress over several hours. However, many studies have consistently shown that with the adrenal activation and the sustained release of cortisol, animals are less active, less efficient, less able to tolerate adversity, and vulnerable to rapid deterioration under adverse conditions (Sapolsky, 1992). The animal requires elevated cortisol to mobilize diverse end organ systems and to restrain them by decreasing its release. This system, like biological systems, is one among others, and under conditions of adversity can be disinhibited.

Glucocorticoids increase CRH gene expression in the central nucleus of the amygdala and lateral bed nucleus of the stria terminalis, while they decrease CRH gene expression in the classical fashion in the parvocellular region of the PVN (Watts and Sanchez-Watts, 1995).

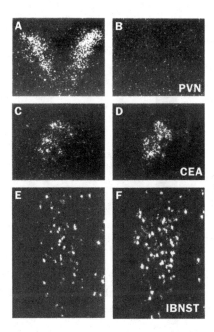

Figure 4.4   Effects of corticosterone in adrenally intact rats.
Corticosterone increases CRH in the CEA (or BNST), while it decreased
it in the PVN (reprinted from Makino, S., Gold, P. W., & Schulkin, J.
(1994a). Corticosterone effects on corticotropin-releasing hormone
mRNA in the central nucleus of the amygdala and the parvocellular
region of the paraventricular nuclues of the hypothalamus. *Brain
Res*, **640**, 105–112, Copyright (1994), with permission from Elsevier.
Reprinted from Makino, S., Gold, P. W., & Schulkin, J. (1994b). Effects
of corticosterone on CRH mRNA and content in the bed nucleus of the
amygdala and the paraventricular nucleus of the hypothalamus. *Brain
Res*, **657**, 141–149, Copyright (1994), with permission from Elsevier).

In fact, even within the PVN CRH neurons that project to the brainstem are not restrained by glucocorticoids, and CRH elsewhere in the
brain are not restrained by glucocorticoids (Figure 4.4).

Implants of corticosterone into the central nucleus of the amygdala not only increases CRH in the central nucleus of the amygdala
(Shepard *et al.*, 2000), but also in the lateral bed nucleus of the stria
terminalis and promotes cautious avoidance and CRH release to noxious stimuli and unfamiliar and fear-related events.

Glucocorticoids also can increase the release of CRH at the level
of the frontal cortex (Merali *et al.*, 2008), and regions of infralimbic

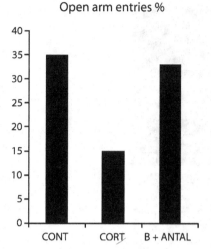

Figure 4.4b   Effects of corticosterone directly implanted into the amygdala on exploration in an open field, with and without a TYPE 1 receptor antagonist (adapted from Myers *et al.*, 2005).

cortex, known for its role in the regulation of extinction learning, is also linked to regulatory functions on PVN CRH expression (Radley *et al.*, 2006). Regions of the prefrontal cortex contain diverse adrenal steroid receptor sites in addition to CRH sites (McEwen, 1995). Bilateral projections to and from the amygdala to parts of the infralimbic cortical regions put both regions in essential contact in the regulation of social contact and avoidance.

Changes in infralimbic cortical regions underlie extinction to fear-related events (Quirk *et al.*, 2000), and may be linked to temperamental differences in social events (Kagan *et al.*, 1988). Infusions of CRH injections are known to influence fear events within these cortical sites (Sullivan and Gratton, 1999), where glucorticoids are known to influence the HPA axis. Regions of the prefrontal cortex that send projections to the PVN regulate the HPA axis, along with the bed nucleus of the stria terminalis (Herman *et al.*, 1998). Regions of the dorsal and ventral regions of medial prefrontal cortex have differential effects on the regulation of PVN CRH. Lesions of the dorsal region enhanced CRH expression in the PVN under duress (Radley *et al.*, 2006).

Importantly, psychological adversity and behavioral adaptation are preferentially regulated by the central nucleus rather than the PVN (Schulkin *et al.*, 2005); one very plausible view is the following. In the regulation of the PVN, negative restraint is tied to vegetative physiological functions, the behavioral coding for which CRH participates as an important information molecule is linked to diverse behaviors, of which fear is one amongst others in mobilizing behavioral adaptations.

Taken together, these findings suggest that a functionally homologous limbic neurocircuitry was present in the earliest tetrapods, thus arguing against the view that large changes occurred in the functional organization of the amygdala in the amphibian-reptilian transition and that the nature of the feedback regulation might have occurred before the divergence of the amphibian and amniote lineages and may be common features in tetrapods.

ALLOSTASIS REGULATION AND THE PHYSIOLOGY OF CHANGE: PREGNANCY AND PARTURITION

One of the defining features in humans is that we have a fascinating 9-month organ: the placenta. Long thought to be just a conduit of nutrients and oxygen from the mother to the fetus, it is in fact an organ that is metabolically active, metabolizing 60–70 percent of the glucose from the uterine circulation (Gluckman and Hanson, 2005).

Importantly, the placenta produces every information molecule that will be produced in every end organ system and, therefore, is quite the endocrine secreting gland. These information molecules include oxytocin, CRH, corticosterone, vitamin D, catecholamines, etc. The placenta is therefore a transformative organ, ancient and vital for the transition to mammals.

We now know that it performs more than the transfer of nutrients as a passive organ, but is a vital 9-month engine of regulation of physiology and long-term changes on brain, behavior and physiology (Power and Schulkin, 2009). In other words, it is a central regulator (mini-brain) of maternal–fetal physiological regulation necessary for reproduction with long-term impact. There is nothing quite like this organ anywhere else in the body, so rich it is in information molecules, although non-placental animals have other routes to facilitate maternal–fetal interactions and fetal development.

CORTISOL INDUCTION OF CRH AND PROFOUND CHANGE BIRTH

Change is most apparent during pregnancy and subsequent delivery; the range of both physiological and behavioral adaptations that permeate pregnancy, delivery, let alone development, is a marvel of nature.

One prominent feature is CRH; by the early 1980s, it was characterized in the placenta, and in normal human pregnancy cortisol and CRH tend become elevated in the plasma over the course of gestation. The expression of CRH and cortisol is greater in the second than the first trimester, and then greatest in the last trimester in humans (Goland *et al.*, 1986)

We know something about several of the functions of information molecules in mammalian birth. Estrogen promotes oxytocin expression, corticosterone promotes CRH expression, and both are importantly involved in parturition (Smith, 2007). CRH is normally associated with the releasing factor from the PVN to the pituitary, but it is a peptide widely distributed in the body with an ancient phylogenetic history. Evolutionary conservation of the role of CRH in mediating developmental changes is suggested by recent demonstrations in mammals that CRH of fetal and/or placental origin determines the length of the gestational period (Smith, 2007). Furthermore, elevations in CRH production under conditions of intrauterine stress may precipitate preterm birth, and placenta CRH may be a unique feature of primate placenta (Power and Schulkin, 2006) (Table 4.7).

Importantly, it was discovered that unlike parvocellular hypothalamic CRH that projects to the pituitary gland (Watts and Sanchez-Watts, 1995), placental CRH is elevated by glucocorticoids (Frim, 1988). During the course of pregnancy in humans, cortisol acts to elevate CRH to promote the normal course of pregnancy (Smith, 2007). Placental tissue treated with glucocorticoid agonists or antagonists either increased or decreased CRH gene expression. Moreover, in several studies, women treated with glucocorticoid agonists, for instance, developed elevated levels of CRH in plasma and in placenta (Frim *et al.*, 1988).

During normal parturition in primates, CRH expression is elevated in placental tissue by 15 weeks, and by weeks 35–40 it is quite high (Frim *et al.*, 1988); one role of cortisol is to facilitate CRH in placental tissue. Diverse in vivo and in vitro studies have demonstrated that glucocorticoid agonists can enhance CRH gene expression in the placenta (Figure 4.5).

This positive induction of CRH in the placenta by cortisol has been linked to cAMP mediated mechanisms; cAMP activates CRH promoter sites (Smith, 2007). This is an intracellular mechanism

Table 4.7. *CRH expression in the placenta of diverse species (Power and Schulkin, 2006)*

Species	Circulating CRH	Circulating CRH-BP
Homo sapiens	Yes; pattern over gestation known	Yes; pattern over gestation known
Pan troglydates	Yes; pattern over gestation known	Yes; pattern over gestation known
Gorilla gorilla	Yes; pattern over gestation known	Yes; pattern over gestation known
Pongo pygmaeus	Unknown	Detected; nonpregnant
Hylobates lar	Unknown	Detected; nonpregnant
Papio Anubis	Yes; pattern over gestation known	Unknown
Papio hamadryas	Yes; pattern over gestation known	Not detected; nonpregnant
Macaca mulatta	Yes; pattern Unknown	Unknown
Macaca silenus	Unknown	Detected; nonpregnant
Mandrill sphinx	Unknown	Not detected; nonpregnant
Ateles sp.	Unknown	Not detected; nonpregnant
Saimiri boliviensis	Unknown	Detected; nonpregnant
Callithrix jacchus	Yes; pattern over gestation known	Detected; nonpregnant
Lemur catta	Not detected	Unknown
Eulemur macaco	Not detected	Unknown
Varicio variegate	Not detected	Not detected; nonpregnant
Rat	Not detected	Not detected
Guinea pig	Not detected	Unknown
Horse	Trace levels in pregnant and nonpregnant	Not detected
Sheep	Not detected	Not detected

preserved across diverse taxa and is phylogenetically ancient. For instance, glucocorticoids both stimulate and inhibit CRH promoter sites in placental tissue (Smith, 2007), and this has been conserved across evolution.

Thus, cortisol is not simply restraining CRH gene expression, but can facilitate its expression and thereby promote a dramatic physiological change: parturition. It may be the case that decreases

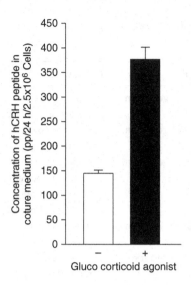

Figure 4.5   Glucocorticoid agonists stimulate CAMP mediated CRH promoter activity in placental tissue (adapted from Nicholson *et al.*, 2004).

in progesterone lead to elevated levels of CRH, perhaps by the greater efficacy of cortisol in receptor sites for this steroid hormone (Smith, 2007).

Importantly, CRH expression and the induction by cortisol are exacerbated under diverse conditions of adversity that could compromise successful reproduction; they include diabetes, preeclampsia, infectious diseases, multiple births, psychosocial adversity and environmental uncertainty during the pregnancy. Interestingly, risk-related behaviors in the mothers during pregnancy have also been linked to elevated CRH and preterm delivery (Figure 4.6).

One hypothesis is that cortisol, by pumping up the levels of CRH before term, is signaling danger (Goland *et al.*, 1986). The normal change in which cortisol and CRH would be facilitated is now exaggerated by the induction and feed forward regulation at the level of the placenta (Frim *et al.*, 1988). But when there is potential danger to both the mother and the fetus, a feature of allostatic load may contribute to the expression of a low birth weight baby. For instance, in marmosets the greater the carry load the greater the expression of CRH (Power *et al.*, 2006); and the greater the metabolic strain the greater the likelihood of preterm delivery. The activation of CRH, in addition to its putative clock-like 9-month mechanism in our species (Smith,

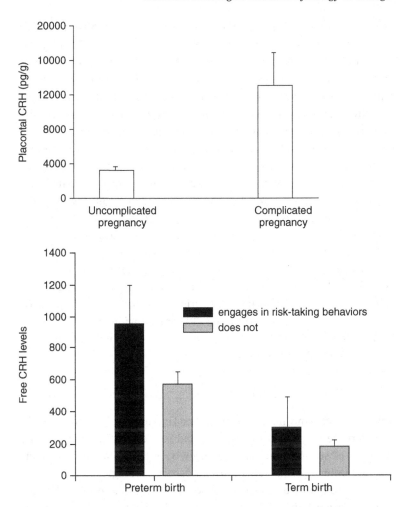

Figure 4.6   Various types of complications in pregnancy result in increased levels of placenta CRH; they include gestational diabetes, preeclampsia, psychosocial adversity (adapted from Goland *et al.*, 1986, 1988). Controlling for other medical factors, risk-taking can still result in elevated CRH and preterm delivery (adapted from Erickson *et al.*, 2001).

2007), promotes fetal movement and maturation by cortisol. In other words, a normal endocrine-related system linked to dramatic change is pushed beyond what it can tolerate, making both the mother and the offspring vulnerable due to allostatic load.

Interestingly, one set of information molecules has one effect, and another quite a different one. To delay parturition is to block cortisol actions. Progesterone is one such mechanism. Progesterone provokes quiescence in both the brain, perhaps through GABA receptors, and in the periphery in diverse end organ systems. That is, the same hormonal mechanism that competes for receptor sites that promotes or prolongs gestation, also acts in the brain to promote action or quiescence (McEwen, 1995). In the brain, GABA is inhibitory and regulated by diverse steroids, one of which is progesterone. In other words, progesterone, a natural antagonist, does the converse and can delay parturition (Smith, 2007); interestingly progesterone, via GABAinergic mechanisms, facilitates a quiescence effect, and in the opposite is true of cortisol. In one case, CRH is elevated and in the other it is not. The mechanisms, in other words, are not restrained by glucocorticoids, but CRH gene expression is facilitated by glucocorticoids – one mechanism that has been suggested with regard to allostatic regulation (Schulkin, 2003). In other words, positive regulation of the CRH gene by glucocorticoids has been described in placental synchtiotrophoblast cells, where rising glucocorticoids of fetal origin upregulate placental CRH, which underlies parturition (Smith, 2007). The positive regulation may be mediated via the CRE, and a region located between -213 and -99 base pairs, which also may mediate cAMP-dependent activation (King and Nicholson, 2007).

METAMORPHOSIS: RADICAL CHANGE, ALLOSTATIC REGULATION

Robert Denver and his group from the University of Michigan have shown quite an interesting parallel and conserved function of CRH; tadpoles of several amphibian species have been shown to accelerate metamorphosis when their ponds dry. Ecological studies suggest that tadpoles use special senses to detect deleterious changes in their larval habitat. Amphibian tadpoles show strong responses to environmental signals, both by expressing multiple, discontinuous phenotypes (i.e. polyphenism) and by exhibiting continuous variation in the timing of developmental events. The most important environmental variable for a tadpole is water availability (Yao et al., 2008). Recent findings show that adrenal steroids and CRH play critical roles in the timing of metamorphosis, as does thyroxine (Denver, 2009). Habitat desiccation results in the precocious activation of the thyroid and the interrenal axes, the hormones that drive metamorphosis (Denver, 2009). Links between several hormonal signaling systems which are that dramatic change (e.g. thyroxine and TRH, adrenal steroids and CRH). With regard to CRH, the CRH signal is transduced by at least two receptor subtypes and may be modulated by a secreted binding protein.

Information molecules play diverse roles depending in which end organ tissue they are expressed, what sites within that tissue, and where they have receptor sites. For instance, CRH facilitates metamorphosis in tadpoles (Denver, 2009), birth in humans (Smith, 2007) augmented by glucocorticoids, and in the brains of mammals CRH facilitates adaptation to changing circumstances (Schulkin, 2003). It is the issue of adapting to change in diverse ecological/social circumstances that underlie concepts of change, such as allostasis. Information molecules, such as CRH or TRH, are bound to changing circumstances; both TRH and CRH in extrahypothalamic sites are not under negative restraints by the respective thyroxine and corticosterone (Thompson *et al.*, 2004); one function of CRH is linked to amygdala function during times of unfamiliar and changing environmental conditions, which is sustained by adrenal steroid activation (Watts and Sanchez-Watts, 1995).

PRE- AND POSTNATAL EVENTS AND CONTINUOUS CHANGE

Diverse prenatal and postnatal events, in addition to genetic makeup, determine the development of the brain. Many events in utero, from maternal nutrition, ecological/social conditions, and medical conditions can impact brain development. But even pregestational events can have long-term effects on behavior; for instance, female rats subjected to various adverse events before pregnancy resulted in a reduction of social behavior in their offspring, particularly in females (Shachar-Dadon *et al.*, 2009). In other words, pregestational adversity (e.g. fear-related events) resulted in decreased social contact and social exploration. Different experiments have demonstrated the profound effect on behavior of prenatal and early postnatal events on long term social behaviors. Across many species, prenatal events help determine sex differences in brain structure and later behavioral functions (intrauterine competition, steroid levels). Yolk itself is a source of steroids (e.g. testosterone) in birds, tracing the impact of steroids and peptides back to egg shell as a resource to be developed (Oftedal, 2002) and utilized.

The hyena is a complex social animal; females dominate and are larger than the males. In utero exposure to testosterone determines intrauterine competition and long-term social dominance (Sapolsky, 1992). Social tracking and group formation is a vital aspect of their social history and adaptation, and these early uterine events set the behavioral trajectory. One intriguing hypothesis based on other literature is that the gonadal steroids set a stage for vasopressin expression, the cephalic expression that underlies the social dominance;

since vasopressin is linked to territorial expression and marking and knotted to the gonadal steroids, this would not be surprising.

Close to my own line of research on CRH, one experiment showed that pregnant animals exposed to adverse events impact the activation of the HPA, and glucocorticoid receptors in diverse regions of the brain, including the hippocampus and extra-hypothalamic CRH, or CRH in their offspring (e.g. Francis *et al.*, 1999). Diverse maternal related behaviors (grooming for instance) can buffer against the over-activation of an HPA axis when later exposed to adversity as an adult, which can be transgenerational (Francis *et al.*, 1999). Thus it is not just the gonadal steroids that have long-term effects on morphology and neuropeptides expression; the adrenal steroids have long-term effects as well (McEwen, 1995).

The activation of cortisol, or corticosterone in rats, impacts the expression of diverse information molecules such as CRH or neuro-transmitter sites in addition to glucocorticoid receptor sites. The events also affect behavior; increased fear-related social behaviors have often been noted from diverse adverse experiences in utero, for example, or following post natal development. These are forms of fetal programming; events in utero set the conditions for long-term vulnerabilities, which affect a range of behaviors from obesity, fearful shyness, preterm low birth weight babies, diabetes and cardiovascular pathology (Barker, 2004), in addition to hippocampal expression (Takahashi *et al.*, 1998). Of course, hippocampal regulation by cortisol secretion and direct application during development can also impact fear-related behaviors. This means that events in utero have long-term effects on brain physiology, morphology and behavior. The other side of evolution is endless adaptation and expansion to limit these adverse responses. In a tradeoff, affected offspring may have more normal health at the beginning of life, versus adverse health-related events at increased risks at the end or middle of life (e.g. fetal origins hypothesis). These events have effects on diverse information molecules (e.g. leptin, neuropeptide Y, dopamine, CRH).

For instance, experimentally altering an enzyme found in licorice (11 b-hydroxysteroid dehydrogenase) transforms corticosteroid receptor sites; one result is elevated levels of CRH gene expression in the central nucleus of the amygdala expression, which amongst other things can affect the level of CRH neurons in the offspring of pregnant rats, perhaps increasing vulnerabilities later in adulthood. By increasing the level of CRH, by glucocorticoid agonists (for instance, altering the MR and GR receptors sites), in the central nucleus of the amygdala in the offspring by the experience of the mother during gestation, there is a greater likelihood in the offspring of cautious avoidance of unfamiliar spaces.

PREDICTION REGULATION AND RESPONDING TO DISTRESS

One advance in the regulation of the internal milieu is the move beyond reactive changes to predictive and anticipatory changes. The diminution of disruption by anticipatory events adds a new level of regulation. HPA function is concerned with reaction to changes, though not exclusively; amygdala function, which contains many of the same regulatory functions to maintain viability, is linked to both the internal milieu and the social or ecological context. Information molecules, such as CRH, are tied to HPA function by the release of ACTH, but also to regions of the brain, such as the central nucleus of the amygdala knotted to the detection of uncertainty or discrepancy (Rosen and Schulkin, 1998).

Negative restraint of the HPA is linked to classical homeostasis; both CRH and vasopressin (Sawchenko, 1984) operate to maintain the same set point and achieve stability. However, endocrine systems, in particular when they involve diverse forms of unpredictability, are about adapting to change and not staying the same; the social context, the ecological niche, and the adaptation results in speeding up endocrine systems (Wingfield, 2004). This is particularly true in social contexts, in which discerning the hierarchy dominance, the social context, access to food, and social contact will determine the endocrine status, in this case cortisol status (Wingfield, 2004). It is not social homeostasis, but something more akin to allostasis, adapting to the change, regulating diverse physiological and behavioral systems with varying set. The regulation of CRH by glucocorticoids is but one example of this social regulation, this allostatic regulation adapting to different contexts. Recall that, like most information molecules (e.g. insulin, CCK, leptin, vasopressin), CRH is distributed across diverse end organ systems (e.g. heart, spleen, intestine) that play diverse roles in the organization of physiological and behavioral viability over a lifetime.

ALIMENTARY AND VISCERAL CONTRIBUTION

Importantly, information molecules like CRH are widely distributed across the alimentary systems. They are represented in the stomach, for example. Distress signals from adversity are conveyed partly from the activation of these information molecules to the brain; peptides such as CRH are activated in diverse sites such as when there is inflammation, or gastrointestinal distress. The activation of CRH in these peripheral sites conveys distress by specific end organ systems to the brain; in the brain, the same CRH peptidergic systems are bound to organization of behavioral responses to ameliorate the distress. High levels of corticosterone that result from distress can maintain and perpetuate it if the corticosterone is not turned down (McEwen, 1995). It can also enhance the visceral responses too; for instance, colonic or other visceral motor related distress. The distress is expressed by the activation of CRH and the brain is informed by this same signaling systems. But the feedback is bidirectional.

Thus, corticosterone infused into the region that surrounds the central nucleus of the amygdala provokes visceral motor distress as the hormone bathes this region of the brain over several days; when tested, colonic inflammation to a challenge is enhanced. One form of relief is chemical blocking of the CRH signaling system; thus antalarmin, a CRH TYPE 1 receptor antagonist, can ameliorate some of the distress (Myers et al, 2005). Visceral distress is a fundamental organizing principle in approach and avoidance behavioral responses. Activating CRH when there is distress in peripheral CRH sites is but one way to generate adaptive behavioral responses. One system in this case from the periphery to the central nervous systems is instrumental in the behavioral adaptation; CRH is obviously not the only system that underlies these complex biological events, but it is one.

CONCLUSION

Information molecules are ancient, and a form of speciation and extension has expanded their functions across end organ systems in the periphery and in the brain. They are widely distributed across organs and, therefore, are used to play roles common across organs and differences towards viability. The list of peptides, steroids, and neurotransmitters is quite striking; and the regulatory mechanisms have both restraining and enhancing capabilities. Thus, while negative restraint is a fundamental feature of limiting their expression from overuse and exhaustion, positive feedback or feedforward allostatic systems underlie the physiology of change. One information molecule that has some common features is CRH, both at the level of the placenta and parturition and extrahypothalamic sites in the brain; in both, cortisol can enhance CRH gene expression and under pathological extremes can invoke adaptive and longer non-adaptive responses (Wadwha *et al.*, 2001).

As we turn to the next chapter, we will discuss the diverse neuroendocrine systems linked to social behavior and regulatory systems, which are feedforward in neural design and perhaps reflect an allostatic mechanism in which anticipation, and not just reaction, is part of the adaptive capabilities. We now know that the brain can be altered at a number of stages in development, pre- and postnatal, and can result in long- and short-term changes of the brain in part through diverse information molecules and their expression. The diverse phylogenetic trees and molecular ancestors compose a wonderful avenue of inquiry, demonstrating origins and radiations of function, similarity retained and differences magnified; the molecular signatures are the building blocks, but selection takes place at more than one level of analysis (Foley, 2006).

The age of molecular biology, replete with knockout mice, has pointed out a fair amount of compensatory responses, which are important to know, and are a very powerful tool in understanding behavior. They do require integration into levels of analysis, structure and function, genetic and epigenetic levels, and we turn to more of this discussion in the next chapter.

5

Social Neuroendocrinology

INTRODUCTION

One feature that underlies cephalic regulation of behavior is the diverse social cues for approach and avoidance. The greater the sophisticated social context, the larger the cognitive/behavioral strategies to draw upon (Byrne and Bates, 2007). Behavioral approach and avoidance are both cognitively and affectively driven; all neural systems are rich in information processing, and therefore cephalic adaptation of all sorts is cognitive in nature. The issue is not really cognitive versus not cognitive, but how flexible, labile and adaptive the systems are (Gallistel, 1980).

In this chapter, I underscore the diverse information molecules that underlie adapting change or allostasis. Many of them reflect feedforward systems: steroids facilitating neuropeptide expression, adapting and coping with social change or social stability (Herbert and Schulkin, 2002). The chapter begins with a theme broached in the last chapter with regard to organization and structural changes in the brain that underlie diverse forms of social behavior, followed by a discussion of appetitive and consummatory motivated behaviors essential for successful social behavioral adaptations. I begin first with a discussion of appetitive and consummatory concepts, followed by describing socially related phenomenon.

APPETITIVE AND CONSUMMATORY BEHAVIORS

Just what are they? An early description by Craig (1918) depicted diverse appetitive approach behaviors, and consummatory behaviors in diverse species; central states of the brain underlie the search

95

engine and the consummatory phase of motivation (Stellar, 1954). The motivation for rewards pervades ecological space. Learning is bound to it: how to acquire something, how hard one will work to attain it, and then of course the consummation. Both appetitive and consummatory behaviors underlie behavioral adaptation. One is the search engine and the other is the consummation of the goal object; the appetitive object is social contact, social stability, social well-being, and consummation is the realization of this (Jaspers, 1913/1997). Placed in both an ethological and a psychobiological perspective, this conception could account for many forms of regulatory behavior (Stellar, 1954). This distinction is also limited since not all forms of behavioral expression are codified from within this perspective; but it is a rough heuristic biological ground to account for a fundamental way in which to understand behavioral adaptation. And it is an important distinction in understanding social behaviors.

In infants, acoustically laden for a receptive care giver, the approach behaviors that initiate behaviors towards consummatory satisfaction and some sort of comfort underlies a broad array of behaviors of infants, of mammals, of birds. The theme is ancient. But it goes two ways, or multiple ways; the external circumstance matters and helps determine the internal appetitive responses. The external availability of the resources redirects the appetitive response; there is a competition of appetitive responses broadcasting the appetitive concerns across the central nervous system in diverse information molecules and across neural circuits (Table 5.1).

CEPHALIC ADAPTATION: FEEDFORWARD SYSTEMS IN THE BRAIN
AND APPETITIVE AND CONSUMMATORY BEHAVIORS

One form of regulation of feedforward systems in the brain is steroid induced, and regulates diverse information molecules (Pfaff, 1980). Feedforward regulatory systems, namely the positive induction of CRH by glucocorticoids, are tied to allostatic regulation. They underlie many appetitive/consummatory behavioral relationships.

Glucocorticoids and extrahypothalamic CRH is just one example amongst others in which steroid hormones regulate neuropeptide gene expression in the regulation of behaviors that serve physiology, such as successful reproduction. In fact, there are both negative and positive feedback systems in the interactions between steroid, hormone, and neuropeptide that underlie behavioral responses; they

Table 5.1. *The essential phases of appetitive and consummatory behaviors (adapted from Swanson, 1988)*

Initiation phase
Deficit signals
Incentive and exteroceptive sensory information
Cognitive information (conditioning, anticipatory)
Circadian influences
Long-term memory

Procurement phase
Arousal (general)
Foraging behavior
Locomotion
Sensory integration
Previous experience
Short-term or long-term memory
Incentives
Maintenance of homeostasis (visceral integration)
External cues

Consummatory phase
Programmed motor responses
Discriminatory factors
Satiety mechanisms
Reinforcement
Hedonic motivation

Competition for behavioral expression
Multiple motivational states
Environmental factors
Assessment of success/failure

range across vertebrates, and include sexual behavior in newts (Moore and Rose, 2002).

While an exaggerated emphasis has been on negative restraint as a primary model for steroid peptide interactions on the model of the HPA axis, the behavioral model is less in conformity with this view.

Let me illustrate with several examples, many of which I have already either discussed or alluded to. Estrogen, for instance, interacts with diverse neuropeptides and neurotransmitter systems (e.g. enkephalins, dopamine). Estrogen can increase oxytocin gene expression under diverse contexts in extra-hypothalamic sites, and is well known to facilitate the induction of oxytocinergic in diverse regions of the brain that underlie sexual motivation. Classical examples abound from the hormonal literature detailing its effects on the brain in the expression of behavior; high levels of estrogen can lower the threshold for diverse stimuli related to sexual contact (Pfaff, 1980).

Steroids (e.g. estrogen) affect arousal systems important for a wide range of behavioral/cognitive adaptations (Pfaff, 1980). Estrogen can increase oxytocin in diverse regions of the brain that underlie behavioral functions that range from sexual function, some of which can be rapid (McEwen, 1995), suggesting both membrane (and genomic) effects, to aggressive behavior in relatively safe environments. Estrogen can promote active exploration and has diverse effects on a number of other neuropeptides. It facilitates prolactin expression and promotes diverse forms of maternal behavior. Estrogen can also increase CRH expression in rats in unfamiliar environments and increase fear-related conditioning (Jasnow et al., 2006), and it underlies the regulation of dopamine, essential for the organization of action; in fact, estrogen is known to reduce fear and increase activity (Richter, 1943). However, it does so contextually; in familiar environments it increases activity and reduces fear, in unfamiliar it does the opposite (Figure 5.1). More generally, estrogen promotes activity (Richter, 1943).

Arousal systems are both specific to a context and knotted to general activation that underlies diverse behaviors; by increasing CRH gene expression in the central amygdala, estrogen facilitates one role of CRH: to attend to changing circumstances to environmental events (Merali et al., 2008).

As I have indicated in Chapter 4, corticosterone is typically known for its restraint on CRH production in neurons that project to the pituitary (Swanson and Simmons, 1989); these are profound effects, as is the general inhibition of diverse inhibitory input on other hypothalamic releasing factors by other steroid hormones. But it is not PVN CRH which is important for the behavioral effects of CRH. It is the amygdala CRH that underlies the behavioral adaptations (Davis et al., 2010).

Other good examples of steroid and neuropeptide interaction are the adrenal steroids and angiotensin; aldosterone and corticosterone increase angiotensin in the brain to increase water ingestion (Epstein, 1991). These steroid hormones are important to sustain the morphology and the induction of these neuropeptide systems in the

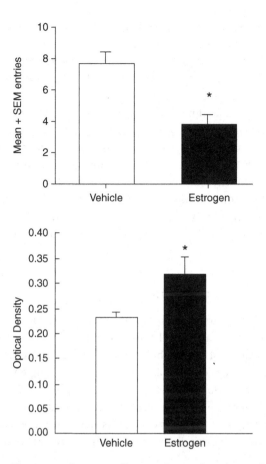

Figure 5.1   Estrogen effects on fear and corticotrophin releasing hormone gene expression in the central nucleus of the amygdala (Jasnow et al., 2006). Mean percent time freezing to a context (Jasnow et al., 2006). Mean CRH mRNA levels in the amygdala from EB-treated and vehicle-treated mice killed 24 hours after treatment. EB treatment significantly increased CRH gene expression in the amygdala 24 hours after treatment, (but not 2 hours after treatment and not after 10 days of treatment) compared with vehicle-treated animals as shown in the figure (reprinted from Jasnow, A. M., Schulkin, J., and Pfaff, D. W. (2006). Estrogen facilitates fear conditioning and increases CRH mRNA expression in the central nucleus of the amygdala. *Hormones and Behavior*, **49**, 197–205. Copyright (2006) with permissions from Elsevier).

brain. The hormones that regulate peripheral balance act in the brain to generate adaptive behaviors towards the same end.

Aldosterone, for example, a sodium-conserving hormone secreted from the adrenal gland, is produced when sodium is needed, to conserve sodium and redistribute it from bone to blood to maintain extra cellular fluid composition. As well, aldosterone facilitates the expression of angiotensin in the brain to generate the behaviors bound to the appetitive behaviors associated with sodium hunger (Epstein, 1991). Vitamin D, the hormone of calcium re-absorption linked to parathyroid hormone (PTH) in the brain, may be linked to calcium on consumption in addition to calcium conservation. Many of these regulatory mechanisms, however, have social functions: social contact and social withdrawal, attention to external events, etc.

SONG AND SOCIAL COMMUNICATION

Organizational and structural events related to hormonal systems are tied to critical periods that facilitate the organization and the later conditions for the activation of diverse behaviors essential for successful adaptive responses. They occur pre- and postnatally. Lifelong change, maintenance, and evolution are features of the brain (McCarthy, 2008), supported by diverse steroids (e.g. estrogen promoting tissue, corticosterone for long periods deteriorating tissue) and by the induction of neuropeptides (growth factors, oxytocin, angiotensin, CRH, vasopressin).

Steroids secreted during pre- and postnatal development have many different effects on the brain. For example, we have known for some time about the defeminization of the brain by the secretion of testosterone during postnatal development (McCarthy, 2008) through an aromatization that is fundamental and that in fact can occur in utero levels. The gonadal steroid hormones impact postnatal development and expression, and that continues to impact behavioral events and the neural conditions that render them possible.

By an aromatization process, for example, testosterone is converted to estrogenic compounds that have long-term structural impact on the brain and on behavior; bird song is but one example. Song is linked to an aromatization in many species of song birds (Ball and Balthazart, 2002). Song plays diverse roles from social calls, warnings about predators, and food resources and mating calls. Song is syntactically structured and formed during critical periods in which acoustic signals are vital in addition to the steroid conversion for the expression for adult adaptive song (Arnold, 2002).

Song, by definition, is social; it calls to others. There is a rich semiotic component; having had the occasion to have taken a class with Peter Marler, one of the leading experts in bird song, many years ago, it always struck me how beautiful song is, as is the syntax that has been understood to be associated with it, and the hormone tied to it. A conversion process that transforms testosterone into an estradiol by an enzymatic process is truly remarkable as a piece of natural history (McCarthy, 2008).

Importantly, the steroid induction of neuropeptide regulation is a common theme; vasotocin, an analogue of vasopressin, is linked to diverse forms of behaviors, including mating-related behaviors and mating calls in diverse species (frogs, Marler et al., 1995), and singing behaviors in the canary (Voorhuis et al., 1991). For instance, vasotocin analogs, at some time periods during the year in the laboratory, are known to enhance the singing behaviors of canaries in testosterone treated animals (Voorhuis et al., 1991)

Thus, testosterone is vitally involved in song production in birds (Marler et al., 1988), starting with its presence in the yolk of maternal testosterone in developing birds (Schwabl, 1993). This may increase vasopressin in diverse regions of the brain that underlie song and sexual and territorial aggression (Voorhuis et al., 1991). Again, the behavioral effects of the steroid effects can be rather rapid via membrane-related events; in other words, the hour-long effects of steroid in gene-related impact can be shortened to minutes via membrane related mechanisms (McEwen, 1995), which impact diverse behavioral responses (Moore and Orchnick, 1991).

While testosterone induction of vasotocin facilitates song production in many bird species (Marler et al., 1995), estrogen also has a profound effect on the brain, with the organizational effects (McCarthy, 2008) that promote changes in structure and function essential for diverse social behaviors. Gonadal steroids then trigger vasotocin or vasopressin, which has diverse effects depending upon the species and context. Vasopressin, like oxytocin, is phylogenetically ancient, and linked to diverse functions including osmotic and fluid regulation (Bentley, 1982). Both vasopressin and oxytocin are richly expressed in the brains of all vertebrates studied to date and date back millions of years (Strand, 1999) (Table 5.2).

One important example is decreasing the activation of testosterone during critical periods of development via vasopressin expression

Table 5.2. *Oxytocin and related peptides*

Oxytocin	Placentals, some marsupials, ratfish (*Hydrolagus colliei*)
Mesotocin	Marsupials, nonmammailian tetrapods, lungfishes
Isotocin	Osteichthyes
Glurnitocin	Skates (Chondrichthyes)
Valitocin	Sharks (Chondrichthyes)
Asvatocin	Sharks (Chondirchthyes)
Phasvatocin	Sharks (Chondirchthyes)
Cephalotocin	*Octopus vulgaris* (molluscs)
Annetocin	*Eisenia foetida* (annelids)
Vasotocin	Nonmammalian vertebrates, cyclostomes
Vasopressin	Mammals
Lysipressin	Pig, some marsupials
Phenypressin	Macropodids (marsupials)
Locupressin	*Locasta migratoria* (insects)
Arg-conopressin	*Conus geographicus* (molluscs)
Lys-conopressin	*Lymnaea stagnatis* (molluscs)

in the brain (DeVries and Miller, 1998). A first point to note is that vasopressin was traditionally associated with pituitary release in the control of water volume at the level of the kidney when water balance is compromised; when water balance is depleted, pituitary vasopressin acts on the loops of Henle within the kidney to facilitate water conservation, to absorb and hold onto water (Fitzsimons, 1979).

However, vasopressin at the level of the kidney is only one function of vasopressin. As a neuropeptide it takes on many diverse roles and (it also plays more than one role in the periphery). One of its roles in several regions is in the maintenance of territorial defense, inducing defense aggression related to external circumstance and sustained by gonadal activation. It is also active in the induction of behavioral competence; for example, the territorial aggression that is required to conserve the physical space. The expression of vasopressin is under steroid regulation, in addition to critical periods setting down conditions of vasopressin in diverse regions of the brain that underlie social behaviors.

Social calling behaviors, of course, are not unique to birds. There are changes in the brain in lower species reflecting hormonal changes (Marler and Hamilton, 1966); the steroid secretion sets the conditions for facilitating structural changes that then are activated across the life cycle. Some of these events occur in utero, from temperature or warmth, which facilitate song production and perception (Nottebohm, 1994).

The different social calls serve diverse roles in different species, and they are almost always rooted in social functions. In many species, gonadal steroid hormones facilitate a wide range of social behaviors. Like language and other forms of animal communication, the calls are semiotic (Marler, 1961), rich in information about territory, social groups, alliances, predation and danger, resources, etc. the life blood is embedded in the diverse cognitive systems for retrieving and signaling information. In a brain prepared at each level of the neural axis to participate in the orchestration of, say, facial expression to the appropriate context, the hooting sounds to the social milieu in which, for example, the chimpanzee finds itself (Marler and Hamilton, 1966).

The social communicative components are widespread (Arnold, 2002). Communicative competence, whether song or other forms of intimate social contact, underlies our evolutionary ascent. One mechanism for this is the regulation of the information molecules in the brain.

One role for aromatization is to induce structural changes in peptides and neurotransmitter levels in the brain (Ball and Balthazart, 2002). And one result is in altering diverse information molecules, such as peptide hormones, that underlie behavioral adaptation; this is not only the structure of particular nuclei, but the information molecules that are produced in diverse regions of the brain. Vasopressin is but one example (De Vries and Miller, 1998).

VASOPRESSIN AND SOCIAL BEHAVIOR

Affiliative behaviors are fundamental to a variety of species, and in some species with long-lasting affiliations, they are linked to differences in neuropeptide expression (oxytocin, vasopressin; Donaldson and Yong, 2008). Slight changes in the vasopressin gene may underlie these differences in social behavior (Figure 5.2).

Steroid hormones, such as estrogen or testosterone, are essential in sustaining the neuropeptide expression and receptors (McEwen, 1995). They underlie the behavioral regulation for approach/avoidance behaviors; vasopressin 1a receptors underlie maternal defensive aggression mediated by estrogen levels (Nephew and Bridges, 2008). Vasopressin also underlies social communicative functions that include human social contact and affiliative behaviors (Thompson et al., 2006). There are several types of vasopressin receptors in which vasopressin type 1a

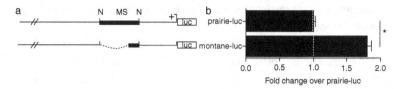

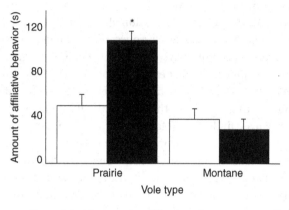

Figure 5.2   Top: species differences in structure at the microsatellite region yield significant fold changes in expression in A7r5 cells. Schematics (a) represent the species constructs tested. N represents the Nde I sites generated on both constructs during cloning of the plasmids, MS represents the large prairie microsatellite and small mountain microsatellite regions. The montage vole construct results in significant fold change over the prairie vole construct. Data are presented as the average of two independent experiments; mean ± S.E.M, n = 6/experiment. Reporter activity of montane-luc compared to praire-luc; * $p < 0.0001$ (Hammock, E. A., & Young, L. J. (2004). Functional microsatellite polymorphism associated with divergent social structure in vole species. *Mol Biol Evol*, **21**, 1057–63, by permission of Oxford University Press). Bottom (b): male monogamous prairie voles, but not promiscuous montane voles, spend significantly longer in affiliative behavior (olfactory investigation and grooming) with a stimulus female after central administration of AVP (filled bars) compared to those subject to administration of saline (open bars) (reprinted by permission from Macmillan Publishers Ltd: Young, L. J., Nilsen, R., Waymire, K. G., MacGregor, G. R., & Insel, T. R. (1999). Increased affiliative response to vasopressin in mice expressing the V1a receptor from a monogamous vole. *Nature*, **400**, 766–8., copyright (1999)).

is linked to diverse social behaviors; for instance, maternal aggression, territory, and social recognition have been consistently linked to this subtype in mice and rats (Nephew and Bridges, 2008). Indeed, central vasopressin is essential for partner preference and aggression in monogamous voles; central vasopressin inhibitors interfere with these behavioral responses. Central infusions of vasopressin into the CSF facilitates partner preference in the monogamous vole; manipulation of the vasopressin receptors can alter the behavior, either resulting in increased or decreased affiliative behaviors (Figure 5.3).

Transgenic manipulations of mice from voles increases affiliative behaviors (Carter *et al.*, 1997/1999) through a vasopressin mediated

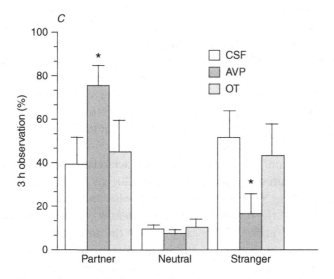

Figure 5.3   Central infusion of AVP increases aggression and induces partner preference. During central infusion of CSF, OT or AVP (0.5 ng h⁻¹), each male was housed with a non-receptive female 'partner' for 24 h. In a subsequent preference test, CSF- and OT-treated males spent equal time with a novel female (stranger) and the partner ($t - 0.51$, NS and 0.17, NS for CSF and OT, respectively). AVP-infused males spent more time with the partner than the stranger ($t = 3.26$, $p = 0.01$), with a significant overall treatment difference in time with the partner ($F(2,22) = 3.3$, $p = 0.05$) attributed to a difference between AVP-and CSF-treated males ($t = 2.38$, $p = 0.05$) (reprinted by permission from Macmillan Publishers Ltd: Winslow, J. T., Hastings, N., Carter, C. S., Harbaugh, C. R., & Insel, T. R. (1993). A role for central vasopressin in pair bonding in monogamous prairie voles. *Nature*, **365**, 545–8, copyright (1993)).

receptor, and variations in maternal care are associated with vasopressin and oxytocin receptors (Francis *et al.*, 2002). For instance, partner affiliative behaviors in transgenic manipulation can be affected by altering a vasopressin receptor (Young *et al.*, 1997); manipulation of the vasopressin 1 receptor results in either decrease or increase of social affiliation. Transgenic manipulations, in which this form of the receptor (V1a) has been compromised, result in degradations of social recognition.

Steroids, again (e.g. vitamin D, aldosterone, corticosterone) are essential in sustaining the expression of the neuropeptide and its receptors. Hormonal signals orchestrated to the behavioral regulation for approach/avoidance behaviors; vasopressin receptors underlie maternal orchestrate aggression impacted by estrogen levels (Nephew and Bridges, 2008), and vasopressin facilitates social communication, which includes human social contact and affiliative behaviors (Thompson *et al.*, 2006). Aberrations of social contact have an impact on this peptide system, in addition to a number of others. The induction of the behaviors reflects the activation of vasopressin related peptide and its receptors. Social scent marking is one set of behaviors linked to vasopressin receptors in the brain that underlies appetitive and consummatory behaviors.

In addition, the greater the approach-related behavior, the higher the expression of oxytocin and CRH receptor sites within the region of the nucleus accumbens (Olazabal and Young, 2006). Of course, this was the region that the great twentieth-century anatomist Walle Nauta identified as the way by which the amygdala was linked to limbic motor output underlying approach and avoidance behaviors. It contains diverse CRH receptors that underlie many different behaviors (Berridge, 2004), perhaps linked to attention to external events and the organization of action.

Thus, infusions of oxytocin, for example, into the brain activate enkephalin receptors (one form of endogenous opioid receptors tied to social reward); the nucleus accumbens, a brain region linked to motivation, is of course tied to social contact and social avoidance (Figure 5.4).

PROLACTIN

Prolactin is a fundamental peptide, like oxytocin, both in the pituitary and other end organ systems. Like most information molecules, it has been sequenced, and an evolutionary trend has been depicted

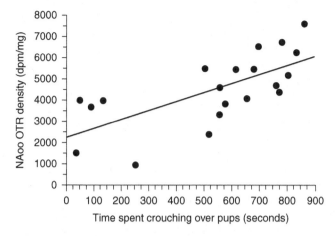

Figure 5.4   Correlation with time spent in attachment behaviors of oxytecin receptor density in the basal ganglia (caudate putamen, nucleus accumbens (reprinted from Lim, M. M. & Young, L. J. (2006). Neuropeptidergic regulation of affiliative behavior and social bonding in animals. *Hormones and Behavior*, **50**, 506–17. Copyright (2006). With permission from Elsevier).

(Wallis *et al.*, 2005). Brain prolactin underlies diverse behavioral events (McCarthy, 2008). Like other fundamental peptide hormones, it is found in diverse vertebrates and invertebrates (Strand, 1999). There are, for instance, several types of prolactin receptor subtypes.

The diverse neuropeptides (e.g. prolactin) overlap across diverse regions of the brain that run from the brainstem to the forebrain; as I have indicated, these information molecules are represented across the brain and many bodily organs participate in their production and regulation in the periphery and the brain. With regard to angiotensin, CRH, prolactin, oxytocin and vasopressin, there are many common sites.

External events help trigger appetitive events by facilitating the expression of different molecules in the brain. The peptide prolactin is one case; long noted for its role in reproductive behaviors it also figures in the "brooding" behaviors of birds. Different forms of this are expressed across a vast array of species and expressions (Tinbergen, 1951/1969). Thus, prolactin in various species plays a wide range of roles in parental behaviors, including brooding and regurgitation, an analogue of pregnancy and lactation in mammals. Depending upon the species of bird, and whether both parents participate in parental behavior, prolactin is playing an important role.

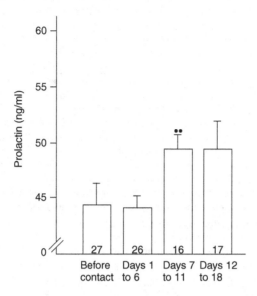

Figure 5.5   Changes in plasma prolactin levels before and during brooding (reprinted from Garcia, V., Jouventin, P., Mauget, R. (1996). Parental care and the prolactin secretion pattern in the king penguin. *Hormones and Behavior*, **30**, 259–65. Copyright (1996), with permission from Elsevier).

Those of us who have kids have been forced to watch animated films such as *Happy Feet*, but these experiences have also reminded us that the lifecycle of the king penguin is a wondrous, beautiful thing to behold; hormones such as prolactin are secreted and tied to chick rearing in both the male and the female (Figure 5.5; Garcia *et al.*, 1996).

Brooding behavior – sitting on eggs till they hatch – is both internally generated and externally realized. In those species in which males and females sit, or perhaps just males, the level of prolactin in the brain helps determine the brooding behavior, but the social milieu can facilitate the release of prolactin and the brooding (Buntin *et al.*, 1992). There are other behaviors related to prolactin levels indicative of social cooperative interactions, such as the regurgitation of food sources (Buntin *et al.*, 1992). Indeed, infusions of prolactin into the brain facilitate these behaviors tied to successful growth in the neonate; diverse forms of helping behaviors in both birds and mammals have been linked to prolactin levels (Schoech *et al.*, 1996).

The level of prolactin is determined, in part, by the social milieu. The appetitive begging behavior of birds helps facilitate, for example, the regurgitation of food products essential for growth. Both sexes, depending upon the species of bird, figure in the regurgitation or feeding of the offspring; central prolactin plays an important role (Buntin et al., 1992), as do estrogen-regulated estrogen receptors (McCarthy, 2008). Both males and females are linked to prolactin stimulated social cooperative behaviors; in the case of retrieval behaviors associated with parental care, expectant females and males displayed greater levels of prolactin (e.g. mice).

There is some evidence that males who respond to a greater degree than other fathers to their offspring have higher levels of prolactin, though experienced fathers may not require prolactin (Roberts et al., 2001; Figure 5.6). Several steroid hormones facilitate prolactin's behavioral responses, and a number of steroids, including cortisol, are essential for the energy required for reproductive behaviors (McCarthy, 2008).

For instance, estrogen facilitates maternal behavior by the induction of prolactin in the brain; estrogen in rats demonstrates both appetitive behaviors (the search for offspring) and consummatory behaviors (suckling of the offspring), or (in the case of the bird) regurgitation of the food source; infusions of prolactin into the sexually dimorphic preoptic nucleus facilitates the behaviors associated with regurgitation (Buntin et al., 1992).

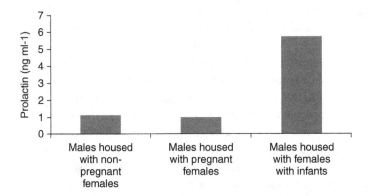

Figure 5.6   Plasma concentrations of prolactin in male marmosets in captive family groups containing, left to right, nonpregnant females, pregnant females, and females with infants aged 10–30 days. Each histogram shows the mean concentration for 10 samples from a single male, except for two of the males caged with pregnant females; six and seven samples were collected from these animals before their female partners gave birth (adapted from Dixson and George, 1982).

A ritualized set of social behaviors may be apparent; the lower the species the more ritualized as a rule, and the more limited the behavioral expression (Tinbergen, 1951/1969). It is an external event coupled with internal levels of prolactin that figures in this important behavior. Steroids figure in setting the condition for prolactin expression in regions of the brain like the hypothalamus and amygdala (Buntin et al., 1992). The central expression role of prolactin then figures in helping to sustain the behavior. In fathers that participate in the caring of the offspring in several species (Figure 5.6; Dixson and George, 1982), levels of prolactin are elevated, but prolactin may not be necessary in experienced marmosets (Almond et al., 2006).

Both fathers and other males, in addition to female marmosets, demonstrate greater levels of prolactin related to physical contact associated with paternal or maternal behavior and care-giving behaviors. This is found in animals ranging from birds to primates to great apes, including humans (Vieck et al., 2000). And, of course, too much is extant in reproductive success to suggest it alone is responsible for these diverse forms of behaviors (Almond et al., 2006).

Depriving the brain of estrogen during development has many effects, one of which is the expression and responsiveness to diverse neuropeptides (e.g. oxytocin, prolactin). For instance estrogen facilitates prolactin-related maternal behavior in experienced rats by the induction of prolactin expression (Bridges and Mann, 1994). These events interact not surprisingly with the experience of the mother, and the social context for the offspring.

Physical contact is associated with prolactin levels, and physical warmth is one important adaptation; prolactin promoting social contact is a fundamental mechanism in species such as our own, in addition to the construction of objects to keep us warm. One reason for the prolactin levels and their elevation is to promote social contact (temperature regulation perhaps for swarms of reptiles such as snakes that can regulate body temperature by contact). One consistent and recurrent result of experimentation over several decades is that, in part, one role for prolactin is in bodily amelioration against adversity (Drago et al., 1989).

Prolactin is also linked to transitional states between daytime and dark periods; like melatonin, it is knotted to dark periods and under circadian control (Wehr et al., 1993), and it is also tied to a wide array of appetitive behaviors, including water consumption. Prolactin itself and its expression influences diverse forms of behavioral expression, including REM sleep; prolactin infusions into the central region of the amygdala facilitate REM sleep (Sanford *et al.*, 1998).

OXYTOCIN AND SOCIAL BEHAVIOR

Oxytocin, as I have indicated, is an amino acid that is phylogenetically ancient; it is linked to vasopressin in its origins and their fundamental links to fluid balance. It is expressed in diverse end organ systems, and there are variants of oxytocin receptors in the brain and other end organ systems. Pre- and postnatal events have long-term effects on oxytocin expression; a number of studies involving macaques have shown that rearing conditions have long-term implications in terms of neuropeptide expression and steroids and neurotransmitters (e.g. CRH, monoamines. adrenal steroids, and of course behavior, Winslow *et al.*, 1993). Classical deprivation experiments are demonstrative: macaques in normal maternal reared conditions developed more elevated levels of oxytocin than the nursery reared macaques with lower levels of CRH; nursery reared animals also have more elevated levels of serotonin (and CRH).

As I indicated in an earlier chapter, oxytocin is tied to milk letdown, osmotic regulation and parturition in the periphery, and social attachment is also a fundamental aspect of this information molecule. Oxytocin infusions into the brain facilitate partner formation (Lim and Young, 2006). Regions of the brain such as the medial amygdala, medial bed nucleus of the stria terminalis and regions of the basal ganglia are importantly bound to attachment behaviors and to oxytocin or vasopressin expression (Young *et al.*, 1997). Many contexts in which social stimuli are evaluated activate oxytocin expression. Indeed, infusions of oxytocin into the shell of the nucleus accumbens facilitate social contact and partner preferences via dopaminergic mechanisms (Liu and Wang, 2003).

One mechanism for social attachment is estrogen receptor regulation and oxytocin expression in the evaluation of social stimuli. Genes encoding for several estrogen receptor subtypes and oxytocin expression and receptors are essential for social affiliative behaviors in mice (Choleris *et al.*, 2006). In addition, genes that code for CRH expression, CRH receptors in the regions of the nucleus accumbens, affect partner preferences in voles (Olazabal and Young, 2006). A putative role for CRH is attention to salient events (Pecina *et al.*, 2006). After all, the world is littered with unfamiliar, potentially dangerous events, contexts in which CRH regulation is vital.

The manipulation of these genes has important effects on approach and avoidance behaviors associated with social contact, but while mice deficient in oxytocin and genetically altered by transgenic methods have many of the mechanisms for mating and giving birth, they are lacking systems for milk ejection and other social behaviors (Carter *et al.*, 1997/1999).

The medial nucleus, a sexually dimorphic region of the amygdala, is tied to olfactory stimulation (Hofer, 1973), and therefore (and not surprisingly), oxytocin infusions into the brain in several species facilitate approach to pleasant odors, and to social stimuli that facilitate approach-related behavioral responses (Febo *et al.*, 2005). In one experiment for instance, oxytocin was shown to underlie the odor preferences of neonatal rat pups (Nelson and Panksepp, 1998). Social memory is tied to oxytocin expression in the medial amygdala (Ferguson *et al.*, 2001). Oxytocin is activated in brain regions to attachment behaviors during suckling in many species (Febo *et al.*, 2005), and also social comforting behaviors (Carter *et al.*, 1997).

For example, experiments using functional magnetic resonance imaging demonstrate that oxytocin is associated with the mother-pup social bonding during suckling (Febo *et al.*, 2005). Key regions of the brain that include the medial preoptic region, medial and cortical amygdala and medial bed nucleus of the stria terminalis, and nucleus accumbens are tied to these events (Febo *et al.*, 2005). Since oxytocin underlies suckling, and therefore is bound to lactation, olfactory transmission is a primary sensory function and infusions of oxytocin promote maternal olfactory associated odors in rat pups (Nelson and Panksepp, 1998).

Oxytocin, as I have indicated at the onset, is a neuropeptide linked to social approach behaviors and is regulated by these events (Carter *et al.*, 1997/1999). In monogamous voles, they increase social attachment and promote diverse forms of parental behaviors in both sexes. Even in male mice with littermates housed in a context where females were expecting to give birth, oxytocin levels were higher (Figure 5.7; Gubernick *et al.*, 1995); in the postpartum period the levels then dropped.

Information molecules, such as oxytocin, play diverse roles depending upon where they are expressed and where they act in the body and in what context. Oxytocin regulated in the brain by steroids, for instance; estrogen-related behavioral responses tend to ameliorate anxious avoidance states (Carter *et al.*, 1997/1999). In another example, induction of central oxytocin in rats tends to reduce the activation of

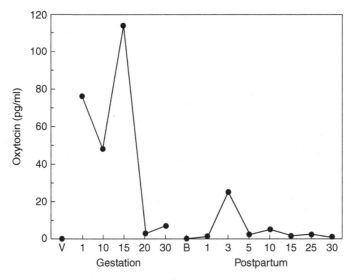

Figure 5.7   Oxytocin levels in male mice when females were expectant and postpartum. Median concentrations of plasma oxytocin in male mice across the reproductive cycle. Blood samples were collected from virgin males (V), expectant fathers throughout the gestational period, and fathers from the day of birth (B) of the litter to weaning. Expectant fathers had significantly higher concentrations of plasma oxytocin than did virgin males or fathers (from D. J. Gubernick, J. T. Winslow, P. Jensen, L. Jeanotte, and J. Bowen, 1995), Oxytocin changes in males over the reproductive cycle in the monogamous, biparental California mouse, Peromyscus californicus. *Hormones and Behavior*, **29**: 59–73. Reprinted by permission).

corticosterone and anxious behaviors (Windle *et al.*, 1997). Estrogen facilitates oxytocin expression (McEwen, 1995). Importantly, one mechanism of estrogen on affiliative behaviors related to maternal care is by the induction of oxytocin. An estrogen circuit within the amygdala is an area linked to social perception and to gathering a quick "gist" of events, a fast and frugal heuristic response that has worked effectively for millennia (Choleris *et al.*, 2006)).

One area of the brain, the medial region of the amygdala, is linked to social perception, innervated by olfactory information. In some of the species studied (mice, voles) it requires the regulation of oxytocin by estrogen. Genomic manipulations of specific estrogen receptors and their interaction with the amygdala impacted social behaviors knotted to olfactory sensibility (Choleris *et al.*, 2006). Again, the medial region is innervated with olfactory input which, in mice,

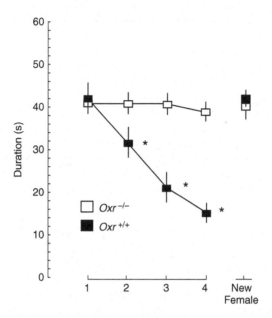

Figure 5.8   Social memory in mice. Social memory by *Oxt-/-* (open
symbols) and *Oxt+/+* (filled symbols) male mice was measured as a
difference in olfactory investigation. Data depict mean +/-1 s.e.m.
for the amount of time(s) allocated to investigation of the same
ovariectomized (OVX) female during each of four successive 1-min
trials. A fifth 'dishabituation' trial depicts the response of males to
the presentation of a new female in a 1-min pairing 10 min after the
fourth trial (reprinted with permission from Macmillan Publishers Ltd:
Ferguson, J.N., Young, L.J., Hearn, E.F., Matzuk, M.M., Insel, T.R.,
Winslow, J.T. (2000). Social amnesia in mice lacking the oxytocin gene.
Nature Genetics, **25**, 284–8. Copyright (2000)).

is vital for diverse social contact. Thus, experiments in which oxytocin expression by knockout manipulations (e.g. oxytocin) have been
tested in social contexts show that subjects did not habituate as readily to familiar conspecifics without the oxytocin gene (e.g. response to
unfamiliar con-specifics, Choleris *et al.*, 2006) (Figure 5.8).

Infusions of oxytocin into the medial amygdala, in particular, restore this basic social function (Ferguson *et al.*, 2000). In other
words, transgenic models in which central oxytocin is compromised
or decreased results in disruption of diverse social behaviors, and thus
transgenic mice in whom central oxytocin is depleted show deficits in
a social recognition test linked to olfactory sensibility. Intraventricular
injections restored this function to normal (Choleris *et al.*, 2006).

In regions of the brain that underlie social regulation, such as the medial amygdala, depletion of the gene through the creation of transgenic models in which oxytocin or estrogen specific sites are depleted in a specific area results in decrements of social behavior. One model is the regulation of oxytocin a gene in the medial amygdala, for instance, by estrogen for social recognition (Choleris *et al.*, 2006). Interference at either the level of estrogen or oxytocin disrupts simple social recognition and memory of social events.

SYMBIOTIC SOCIAL BEHAVIORS, SENSORY SYSTEMS AND EVOLUTION

In our species, and indeed in all mammals, first steps are social, contact with mother, contact with others. These fundamental attachments trigger diverse forms of rich information processing in many sensory systems that contribute in this formation; olfaction and gustatory systems are two particularly important sources of information early on, but vision and audition also play a part. The mother's linguistic expression ("motherese") is a significant, affectively rich source of information for the neonate. The regulation is not homeostatic between mother and baby during this time (though it has been characterized that way), but rather is regulatory. Metabolic and social contacts are crucial events; the fluids that are retained, and the comfort that is afforded the anticipatory behaviors for longer-term viability, are apparent across a wide array of mammals and animals in general.

The social communicative behaviors, the responses to acoustic signals or particularly olfactory and gustatory signals are prepotent information latent with a rich communicative context of social modulation (Insel, 1992); for instance, the quiescence of behaviors after the appetitive calls are mediated by diverse information molecules. Social attachment, of course, has real effects on internal hormonal systems. For example, growth hormone declines and cortisol increases with decreased social attachment (Kuhn *et al.*, 1990); social context influences a number of neuropeptides (CRH, vasopressin, oxytocin, etc.; Ruscio *et al.*, 2007).

For olfactory dependent animals, the smells are prepotent with specific semiotic aromas; a social regulatory system evolved in mammals, such as our species in the regulation of diverse metabolic and social needs.

Separation is too early and too often the bane of existence of the socially dependent animal. The social regulatory interactions of parent and infant, whether suckling or regurgitating food source, as in birds, are primary biological events. The hormones of lactation, oxytocin and prolactin, play diverse roles in both the physiological and the behavioral regulation essential for development and for social attachment.

The expansion of olfactory and gustatory cortex and limbic regions are tied to the signals of small-brained mammals and to a devolution of cortex devoted to olfaction (Curley and Keverne, 2005); but these events vary with the social contexts and the degree to which social learning plays an important part in their regulatory functions.

As we moved away from a narrow universe of information molecules and sensory signals to the larger cultural milieu, our absolute dependence on them diminished. Just as temperature can determine sex, there are diverse forms of sexual reproduction, including sexual transformation and the induction of sexual dimorphic structures and functions in the brain. Our species is not strictly determined by these hormonal signals. They certainly do set the conditions for biological organization, and fundamental events (milk letdown, ejaculation, etc.) cannot happen without them. But we are social animals, and that is what expanded in us; our penchant for learning and experience is embedded in these ancient molecules (Curley and Keverne, 2005).

Three sexually dimorphic nuclei figure importantly in prosocial parental behaviors; they include the medial nucleus of the amygdala, medial bed nucleus of the stria terminalis and the medial region of the hypothalamus (Choleris et al., 2006). All three are different aspects of hormone-related parental behaviors; they control behaviors for lower species, but we also participate. While these sexually hormonal systems underlie core social contact, the evolution of our species is an "emancipation" of this strong link; the strong olfactory dependence from an evolutionary perspective corresponds to our corticalization of function and expansion of visual systems. Interestingly, the size of the sexually dimorphic region of the preoptic region is significantly smaller relative to our brain (Figure 5.9; Keverne, 2004). Our species' dependence on these regions and the hormonal fluctuations reduces with encephalization of function.

No doubt the primary olfactory systems terminating directly into regions such as the medial amygdala are important for diverse forms of social contact, having little to do with reproductive success. However, they are perhaps importantly linked to social approach and avoidance systems essential for longer-term viability. Newborns are prepared to readily associate objects to approach and avoidance.

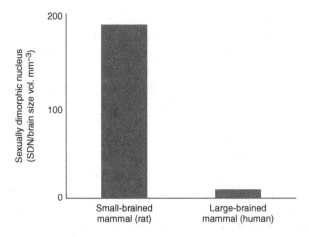

Figure 5.9   Decrease in the primacy of olfaction and corticalization of function in humans (adapted from Keverne, E. B. (2004). Brain evolution, chemosensory processing, and behavior. *Nutr Rev*, **62**, S218–23; discussion S224–41).

Learning is prepotent and selected; core regions of the brain, such as the amygdala, and the activation of information molecules, such as CRH or oxytocin, are linked to diverse forms of learning (Carter *et al.*, 1997/1999). Hormones such as oxytocin and prolactin facilitate the social attachment and the weaning process.

EPIGENETIC EVENTS AND SOCIAL CONTACT

Interestingly, oxytocin is also altered by maternal care across generations of offspring. Cross fostering studies in rodents have shown that variation in maternal care is transmitted in both genomic and non-genomic mechanisms; individual differences in maternal behavior are transmitted from one generation to another (Meaney, 2001). We know that steroids can have both membrane and traditional genomic effects on behavior, including both rapid and delayed behavioral effects (McEwen, 1995), while both mineralocorticoid in addition to corticosteroid play diverse roles.

One example is the link between maternal licking and grooming (high or low), which is consistently transmitted to female offspring; moreover, decreased social comforting contact has long-term implications for most mammals studied. There is wide variation in this phenomenon that has long-term implications on cephalic systems; those rat pups comforted by social contact have greater regulatory capacity as adults on diverse systems in the brain, including neuropeptide and neurotransmitter systems (dopamine, serotonin, CRH).

The regulation of oxytocin and social contact has been linked to an old controversy in biology. Epigenetics is still controversial, recalling debates between Lamark and Darwin about acquired traits and evolution, but has grown in acceptance with the onset of the molecular integration into broader based and regulatory biology. Changes in gene expression without altering the underlying DNA are a workable definition of epigenetics, and are perhaps at the heart of phenotypic alterations in development contingent upon context and circumstance.

David Crews (2008) rightly emphasized a distinction between a molecular and molar level of analysis. The molar level is a critical level of analysis with regard to epigenesis when the consideration is behavioral adaptation and towards an understanding of development. Behavioral adaptations are long noted to be fundamental to evolutionary change, and indeed underlie any definition of evolutionary change (James, 1890/1952). One mechanism that may underlie epigenetic gene regulation is methylization and demethylization (e.g. silencing or enhancing the expression of genes); demethylization prevents transcriptional expression, and one result is the silencing of gene expression. In other words, this represents a putative mechanism for silencing genes in development (Keverne and Curley, 2008).

Variations in parental behaviors, for instance, are linked to oxytocin (or prolactin) expression (Carter, 2007). The high levels of grooming in cross fostering experiments are knotted to estrogen and oxytocin expression (Meaney, 2001), as well as glucocorticoid receptors and diverse neurotransmitter and neuropeptide expression such as CRH, 5HT, dopamine and DNA methylization (Weaver et al., 2004). In contemporary terms, research is showing both epigenetic and genetic contributions to group formation and regulation and ultimately to reproductive success (Keverne and Curley, 2008). Genes that produce oxytocin expression (oxytocin being an important peptide hormone that plays diverse regulatory roles from milk lactation, to parturition, to social attachement) are particularly significant; the same hormone, depending upon where it is expressed in end organ systems, thus plays a wide variety of roles (Carter et al., 1997/1999).

One model alluded to earlier is the transgenic effect of high and low mother infant interactions in terms of social contact through licking behaviors. These events in rodent studies are highly related to olfactory input in the brain, and deep genomic imprinting through olfactory learning is knotted to sexual reproductive success in both sexes (Keverne and Curley, 2008). Importantly, changes in demethylation are linked to the transmission of these social behaviors to the offspring during development in cross fostering experiments. Indeed, manipulations of demethylization impact the transmission of this social behavior and can be reversed later in life (Weaver et al., 2004). The imprinted genes are not permanent; the emphasis on silencing gene expression by imprinting or emphasizing is, of course, an active area of research.

WITHDRAWAL OR AVOIDANCE BEHAVIORS AND CRH

There are subtle differences with these models in rodent social behaviors. One lesson that should be noted and has been learned with molecular science is the danger of deleting one gene, or over-emphasizing one level of analysis or one particular information molecule. Take out a particular CRH gene and one can still note fear-related behavioral responses (Mazjoub et al., 1999). Uncertain comforting social contact and lower levels of social grooming results, for example, in higher levels of CRH in several species that have been studied to date (macaques, rats, mice; Erickson et al., 2005). Developmental placation by robust social grooming and social comfort places the activation of the HPA axis under basal conditions (Meaney, 2001).

What we know is that the postnatal period, in addition to the neonatal, can have a long-term impact in terms of CRH levels and other neuropeptide systems in the brain. One feature impacting these systems is social contact (Carter, 2007). CRH, a neuropeptide linked to social withdrawal, impacts long-term behavioral consequences; animals are more fearful of unfamiliar uncertain events, many of which are social, resulting in greater amygdala activation, higher cortisol levels and greater CRH expression (Kalin et al., 1998).

The medial region, strongly innervated by the olfactory bulb, sexually dimorphic in morphology, and the repository of a number of gonadal and adrenal steroid receptors, is linked to a number of behaviors, including dimorphic sexual behaviors (Neuman, 2008), mineral appetites (Schulkin, 1991) and fear-related behaviors bound to olfactory cues; predator odor, for instance, induces freezing behavior (Takashi et al., 2005). Fear-related predator odors activate diverse gene products in the medial amygdala (Rosen, 2005) in addition to other brain regions; one gene product is CRH. There is a prepotent sensory system designed to detect predators that are innately organized and somewhat separate from other fear-related neural circuits; for instance, diverse cat and ferret odors induce freezing behaviors in rats (Takahashi et al., 2005). The fear-related behaviors are dose dependent; the greater the odor, the greater the fear induced freezing (Rosen, 2005). In the freezing behavior, diverse gene products are induced, including CRH, vasopressin, and various neurotransmitters.

This is an innate activation by the ecological context; the odor results in the activation of corticosterone, which serves to maintain the fear-related response. But while corticosterone-treated rats increase their conditioned fear-related behaviors and decrease their behavioral response to unfamiliar events (Shepard et al., 2000) via the activation of central CRH, they do not seem to affect this innate behavior to potential predators – though CRH is clearly induced by predator odor in regions of the brain such as the amygdala.

Greater avoidance is one result of an innate predatory odor that relates to fear; one function of CRH – though this is speculative – is perhaps to act as a transduction mechanism from an external event to an eternal via the olfactory sensibility. One scenario might be the following; the smell of the odor activates this primary sensory pathway, analogous to the sight of the dog in vision. The effects are immediate via the induction of CRH systems from the periphery to the central nervous system, in this case the medial amygdala. The result is cautious avoidance behaviors.

The sight of a frightening predator or competitor can, within seconds, induce CRH release from the central nucleus of the amygdala. In experiments with sheep, using a dialysis probe to sample the chemical milieu bathing the amygdala, CRH is released at the sight or smell of an unfamiliar dog. CRH is clearly a signal about danger, and amongst other things in the brain, mobilizes action; if at a water spout or a feeding place, the animal would flee. To sustain the fear, the adrenal gland is activated in the mobilization of coherent action, and in this regard the CRH release is sustained. In other words, the initial release of CRH is independent of the adrenal gland hormone cortisol; the second release is cortisol dependent, since blocking its release reduces CRH expression in the amygdala.

In fact, the induction of CRH by cortisol in the amygdala is an important adaptation. There are a number of experiments that have shown that corticosterone can facilitate the induction of CRH gene expression in several regions of the brain; they include the central nucleus of the amygdala and the bed nucleus of the stria terminalis (Makino et al., 1994; Watts and Sanchez-Watts, 1995). Diverse kinds of objects can elicit elevated CRH and diverse neuropeptides or neurotransmitters (bombesin, serotonin, dopamine; Merali *et al.*, 2008) in several regions of the brain, not just fear-related events (e.g. antici-pation of food).

The bed nucleus of the stria terminalis is richly interconnected with the amygdala. The amygdala interacts directly with the bed nucleus of the stria terminalis via regions of the bed nucleus which directly project to the PVN to regulate the HPA axis; more generally, regions of the bed nucleus of the stria terminalis are known to influ-ence HPA function. And some regions within the PVN itself are tied to both anticipatory and reactive forms of regulatory functions (Herman *et al.*, 2003). The BNST may be linked to anxious/ aberrant behaviors through CRH expression (Davis *et al.*, 1997) in which GABA influences HPA functions.

Moreover, all three areas of the brain over-express CRH under diverse conditions (Koob and LeMoal, 2005), including ingestion and withdrawal from various drugs and food. Under these conditions, the adaptive significance of increasing one set of CRH neurons to pay atten-tion and turn down corticosterone to decrease metabolic function has been undermined; it is under these conditions that the system over time deteriorates (McEwen, 1995). The withdrawal from diverse drugs contributes to allostatic changes when pushed beyond physiological/ behavioral viability and adaptation (Koob and LeMoal, 2005).

CONCLUSION

Our neuro-scientific regulation is knotted to others; a set of neuroendo-crine mechanisms orient us towards others. Oxytocin, the hormone of milk letdown and other functions in the periphery (including osmotic regulation) is important for social attachment and social engagement in many species.

The organization of appetitive and consummatory behaviors do not always map neatly to the organization of action, but they account for a wide range of behaviors, including many forms of human social behavior (Tinbergen, 1951/1969). Indeed, a wide variety of interac-tions between information molecules underlie adaptation in cephalic systems, one of which is diverse feedforward systems. These diverse systems underlie regulatory processes and long-term viability through mechanisms perhaps closer to allostasis than traditional homeostatic reactive responses (Sterling and Eyer, 1988).

Social contact is an essential behavioral adaptation for us and kindred species. The profound effect on physiological systems is

Table 5.3. *Paradigmatic examples of steroid effects on neuropeptide expression and behavior (from Schulkin, 1999)*

Steroid hormone	Neural peptides	Behavior
Estrogen Progesterone $\longrightarrow$	Oxytocin LHRH	$\longrightarrow$ Increase female sexual behavior
Testosterone Estrogen $\longrightarrow$	Vasopressin	$\longrightarrow$ Increase flank marking, aggression, territorial defense
Testosterone $\longrightarrow$	Tyrosine hydroxylase (converting enzyme) Substance P	$\longrightarrow$ Increase male sexual behavior
Estrogen Testosterone $\longrightarrow$	Vasotocin	$\longrightarrow$ Increase sex behavior (amphibians), bird song
Estrogen Testosterone $\longrightarrow$	Prolactin Oxytocin Vasopressin	$\longrightarrow$ Increase parental and attachment behaviors
Aldosterone Corticosterone $\longrightarrow$	Angiotensin	$\longrightarrow$ Increase water and sodium appetite
Corticosterone $\longrightarrow$	Corticotropin-releasing hormone	$\nearrow$ Increase water and sodium appetite $\searrow$ Increase fear, anxiety, and depression
Corticosterone $\longrightarrow$	Neuropeptide Y	$\longrightarrow$ Increase food intake

apparent; the relationships are many layered, behavior serves physiology and physiology serves behavior. Cephalic anticipatory systems underlie many of the steroid neuropeptide relationships. Thus importantly, under diverse conditions, steroids facilitate neuropeptide expression that underlies many forms of adaptive behaviors serving both short- and longer-term viability. There is a great range of behaviors that reflect steroid induction of neuropeptide systems and neurotransmitter systems in cephalic systems; the neuropeptides tend to be more specific for the behaviors and neurotransmitters for the background adaptive capabilities (Table 5.3).

Both pre- and postnatal events have both short- and longer-lasting effects on cephalic responses and information molecular and morphological features; environmental events appear to have epigenetic regulatory consequences. The end point is adaptive behaviors, but also omnipresent is devolution of function.

6

Cephalic Adaptation: Incentives and Devolution

INTRODUCTION

We live in a social world with exorbitant and ever-present rewards: from chocolates to stocks, from achievement to thermal comfort. The search for reward is a constant occupation. It is not very surprising that for centuries the concept of reward and human occupation have gone hand in hand in defining our existence, and of course, more than one species is characterized by the search for satisfaction and the avoidance of what is not pleasing.

Diverse theories about pleasure-seeking and pain-avoidance have dominated our intellectual landscape. Hedonism, in both its ancient and modern forms, describes the endless human search for what feels good and the avoidance of what does not. In the broadest sense, hedonic theories are not wrong when construed in terms of the search for satisfaction; but narrowly construed, on the model of a piece of chocolate, the theory does not characterize the full range of our behavioral dispositions, nor that of other species.

Abstract atoms of sensation cauterized from classical hedonism through David Hume (1984/1739) have never done full conceptual justice to the concept of objects. It is objects that orient us, not simple sensation. We are rooted in a world of objects. It is the sight of the bear that renders us afraid, setting off the transduction mechanisms in a well-designed cephalic system to detect and avoid danger. We are oriented towards the object of fear, not its sensation. That is not to deny sensation, for surely that would be a false claim. It is just that, at the level of the world in which we are foraging, it is objects with which we have transactions.

The expansion of the possible range of rewards in the modern world reflects the greater range of objects with which we interact;

with that growth is an expansion of what we might also be vulnerable towards. Consider, for instance, the expansion of obesity; we are exposed to a huge range of different food resources with little actual content, but which tap our vulnerability to the flavors we like. This has led to a near-epidemic of obesity, and an expansion of Type II diabetes. Our reward mechanisms have, in effect, devolved; growth does not always mean improvement, whether biological or cultural.

Addiction is another example of the devolution of the reward mechanism. Addictive objects are endlessly seductive, constantly elevating our ability to obtain satisfaction in decreasing proportion to the object of satisfaction. Like the obesity trap, addiction taps into rewards designed by evolution for meaningful satisfaction. This is not to create a mythology of any original condition, and need not underestimate the elasticity of the reinforcement systems to widening domains. But we must recognize the psychobiological contexts in which the lure of the satisfaction degrades the regulatory cephalic systems (Sterling, 2004).

An expansion of social desires amidst neurobiological systems rich in expectation mechanisms (Schultz, 2007) can create satiation in allostatic systems designed for greater adaptation and elasticity, but the satisfaction may be short lived and misleading (Sterling, 2004). Expectations of reinforcements and the lack of them dominate the conceptual space of normal and pathological addictive states, to which many neurotransmitters (e.g. dopamine, serotonin GABA) and neuropeptides (e.g. CRH, opioid peptides) have been linked (Koob and LeMoal, 2005). A partial list of neurotransmitters linked to both appetitive, consummatory, and withdrawal states from drugs are depicted below.

In this chapter, I discuss the adaptation and the regulation of the internal milieu, simple hedonic shifts and cephalic systems that underlie the maintenance of physiological viability, core feedforward endocrine systems that are the substratum of behavioral adaptation and responsive to incentives and vulnerability to devolution of function. I begin with some core adaptive behavioral expressions, and then shift to devolution of function and the roles of two information molecules: a neurotransmitter, dopamine, and CRH, a neuropeptide. Both have been linked to adaptive and non-adaptive behaviors.

ADAPTATION AND SIMPLE HEDONIC SHIFTS RELATED
TO THE INTERNAL MILIEU

One fundamental mechanism in the orientation towards objects, particularly those related to the internal milieu, is the ability to evaluate what they can provide relative to biological needs: requirements of food resources with proteins and carbohydrates, minerals, fluids, etc. We search for objects related to the satisfaction of biological needs. Depending upon the species, the ecological niche, and the cognitive and behavioral range, there are different forms of adaptation; but the evaluation of food resources dominates much animal behavior.

Animals come prepared to recognize objects and their signifi-cance. One classical example is the ability to learn to ingest salt when not in sodium need (Krieckhaus and Wolf, 1968). Whatever time of day this learning may have occurred and whatever might be associated with subsequent need for sodium, minimal exposure is necessary for this learning.

Another example is the ability to learn the connection between ingestion of a food resource and gastrointestinal distress. Curt Richter (1943) called this phenomenon "bait shyness", meaning foods that are safe and food resources that are not safe, familiar and unfamiliar dimensions. Richter discovered that one example of a toxic substance was enough to decrease a rat's ingestion in subsequent exposures to the food source, a phenomenon that came to be known as taste aver-sion learning.

Moreover, rats learn, with minimal exposure and over long peri-ods of time, relationships between a food resource and visceral ill-ness. The evaluation of this event results in the avoidance of food. But is there devaluation in its hedonic assessment? The event not only affects food ingestion and the evaluation of a food resource, but ceph-alic regulation; insulin secretion, which is known to be released to absorb food resources by gustatory mechanisms or by "conditional associations" as Pavlov put it, is decreased by taste aversion learning (Berridge et al., 1981).

These events can be locked to the time; anticipatory secretion of insulin sets the stage for the absorption and utilization of food resources (Woods et al., 1970). In fact, a wide variety of the information molecules essential for the utilization of diverse energy and mineral resources are released in anticipation to the resources being found and ingested (Denton, 1982). Palatability judgments underlie the for-aging of eatable objects; ingestion of food sources are only partially related to needed resources. Opportunistic ingestion predominates in the real ecological world, but aversive stimuli can change its fea-tures under conditions of need; sea water, aversive and harmful to animals when ingested and not rejected, is ingested with enthusiasm under conditions of body fluid and sodium depletion. In other words, aversive stimuli are now appreciated when found and sought after because of need (Denton, 2005). Both appetitive (the search for the object) and consummatory (finding the object) functions underlie the motivational change (e.g. Tinbergen, 1951/1969).

In fact, there is a predilection in many species, including our own, to like a modest amount of salt; these species are inclined to ingest salt whether it is needed or not (Richter, 1943). In an age when salt consumption is elevated and aggravated and co morbid with other factors that increase our vulnerability to ill health, it is not surprising that a predisposed form of behavior can be exaggerated in expression (Epstein, 1991).

A range of taxonomic facial responses to diverse taste stimuli (sucrose, quinine, salt) are conserved in our evolutionary history. These gustatory signaling systems are important sources of information laden with affective value and utilized later when needed substances might be of relevance (Krieckhaus and Wolf, 1968). In the language of the ecological psychologist J.J. Gibson (1979), an "affordance" – an object of biological significance – is easily encoded by cephalic mechanisms.

Basic gustatory responses are brainstem driven and richly expressed across diverse mammals (Berridge, 2004). The basic liking and disliking of a gustatory stimulus is shown in Figure 6.1. This hedonic approach and hedonic aversion are phylogenetically ancient and underlie the organization of action.

These gustatory or olfactory stimuli generate approach-related behaviors (Stellar, 1954) that are knotted to diverse forms of learning; they also can generate flexible motivational responses. In fact, a key

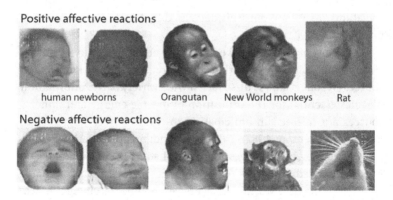

Figure 6.1   Facial expressions in diverse species. Examples of affective facial expressions to taste by human infants, apes, monkeys, and rats (adapted from Berridge, 2004).

issue in motivational systems for us is the flexibility of the diverse forms of operational and cognitive tasks that we can perform. Hedonic shifts permeate the appreciation of commodities that are knotted to regulation, and the context is extended by the social milieu. Those objects associated with sodium (at concentrations that are aversive under normal non-deficient conditions), will be ingested when rats are sodium hungry (Berridge, 2004). This behavior is also known to influence regions of the basal ganglia linked to appetitive behavior. Importantly, incentive salience occurs when the universe of sodium-related objects expand.

Regions of the basal ganglia rich in dopamine expression, the nucleus accumbens regions long thought to integrate motivation and action (Berridge, 2007), are altered by motivational events; sodium hunger or amphetamine craving alter the structure of neurons within this region (Roitman *et al.*, 2002). Predictors of incentives such as sodium or amphetamine alter the neuronal profile with expectations of reward in the ventral pallidal neurons (Berridge, 2004).

Incentive motivations also line the walls of our social interactions, and have large effects on behavior (Bindra, 1974). Liking is not the same thing as wanting something or needing it (Berridge, 2007). Neural systems separate the two. One can obviously like something and not want it; one can want something and not really like doing it. Understanding reward (and the endless seduction of junk food and the vulnerability to diverse modern pathologies, such as obesity) requires understanding of incentives or not wanting to do it (e.g. addicts not wanting to use drugs, and not getting high from them anyway). These can all become major transgressions. An expansion of incentives implodes our social ecological space, making it difficult to diminish the impact of objects of desire.

DOPAMINE, INCENTIVES AND REWARD

Dopamine, more than other broad neurotransmitters, has been linked to reward. But dopamine is an ancient information molecule, whose history transcends vertebrates and has evolutionary multiple receptor subtypes, found in both the peripheral and central nervous system. Dopamine is a primary neurotransmitter in the brain, as is neurepinephrine and epinephrine solely in adrenal modular cells. A conversion process from tyrosine is fundamental in the conversion to dopamine. Dopamine is also a precursor in the metabolic conversion to norepinephrine; one is essential for the organization of action and cognition, the other attention.

Dopamine is the one neurotransmitter that is linked to reward. Dopamine is also fundamental to the processes that underlie the organization of action and the organization of thought. Reward is tied to motivation; motivation requires attention to the external world for the needed resources, and to organization of action to acquire the desired object. But the concept of reward is very complicated; it is not simply one act, and defining the term gets circular quickly. It is not simply the activation of dopamine as was once thought, nor simply an assembly of transmitters or neuropeptides. Nonetheless, reward is fundamental to our lexicon and our understanding of the organization of action and the devolution of function with regard to addictions and afflictions. Reward, I suggest, should function as a heuristic and certainly not as one "thing."

Glucocorticoids, themselves the hormone of energy balance, play diverse roles by inhibiting or restraining in some tissue, facilitating expression in others, serving physiological and behavioral viability (Sapolsky, 1992). In the short term, this is an adaptation, but in the longer-term it increases a vulnerability to pathology. Glucocorticoids may influence diverse reward systems, resulting in less and less satisfaction and more mobilization of action. But they do so in the natural world amidst bouts of plenty and bouts of degraded resources, typical of the ecology of resource allocation other than success, and success varies with the species and the life circumstances. Variation in ontogeny and seasonal variation figure importantly in species allocation of resources. Allostatic regulation favors recognition of diverse forms of cognitive adaptation in managing the internal milieu.

Since dopamine underlies the organization of approach and avoidance systems, one mechanism of regulation is the diverse steroids (estrogen, testosterone, corticosterone, etc.) that enter the brain and regulation dopamine expression. In other words, steroids that promote adaptive behaviors regulate dopamine expression. For instance, the gonadal steroids, by regulating the production of dopamine, increase the dopaminergic neurons tied to reward prediction and incentives, in addition to motivational muscle, to overcome obstacles in order to gain access to mates (Pfaff, 1999).

Dopamine is simply not the hormone of movement, though Parkinson's disease remains fundamentally linked to this information molecule. It also underlies most thought, including syntax in language; Parkinson's patients are often less compromised on the semantic than on the syntactical relationships of language use. It is the memory components that are impaired. The structural syntax by which language is used is in devolution of function as a result of dopamine loss. Other perceptual systems that underlie the organization of action are compromised under dopamine depletion; for instance, the perception of fearful faces and other facial expressions are compromised in patients with low central dopamine. Restoring dopamine levels returns patients to simple perceptual tasks; thus, in fMRI studies in which the amygdala is activated by fear-related depictions, restoring dopamine levels to normal returned the competence of the perceptual response in responding to facial expressions (Tessitore *et al.*, 2002; Figure 6.2).

Regions of the brain tied to the basal ganglia traditionally control motor function, but we now know they affect more diverse functions, including hedonic assessment and incentive salience (Berridge, 2007). For example, infusions of endorphins, which increase the ingestion of sucrose or the liking of it, are more centrally located in the rostral medial region in the accumbens (Pecina *et al.*, 2006). Again, this region has been originally thought to be one anatomical route by which the amygdala influences limbic motor output more generally and where dopamine expression is altered (Berridge, 2007).

Elephants are known, for example, to explore salt licks. Salt licks are places in which diverse salts can be found; one adaptation is perhaps to group a variety of salts in one location (Denton, 1982).

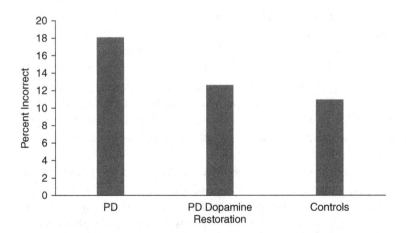

Figure 6.2   Percent incorrect responses in a perceptual detection test in Parkinson's patients (PD), patients who have restored dopamine, and in controls (adapted from Tessitore *et al.* (2002). Copyright © 2002 by the Society for Neuroscience. Reprinted with permission).

A shift in motivation occurs when they are needed (Richter, 1943). Dopamine no doubt underlies organization of action, the sensory/motor cognitive capacity to remember where the resources are and return there to consume the sodium (Krieckhaus and Wolf, 1968). Sodium, in this case, is adaptive; it can also not be, but whether adaptive or not, central dopamine is influenced. For instance, amphetamine and other addictive substances or adaptive sodium hunger both influence dopaminergic neurons within the accumbens; an expansion of neurons within this area is linked to drug craving, not surprisingly, since cognitive/motor functions in a search towards consumption are activated.

And while we can separate liking from wanting a particular reward, and the motivation to labor for it versus sensing it in the environment (Berridge, 2007), all of these are bound to the organization of action, of which reward-related events are a fundamental feature. Indeed, a number of studies show that drug ingestion influences dopamine expression of the shell core of the nucleus accumbens (Koob and LeMoal, 2005). Regions of the basal ganglia may in fact reflect rostro-caudal dimensions, in which a wide array of behavioral expressions are regulated, and in which dopamine plays a fundamental role in the incentive salience and the wanting of eatable objects that are needed by bodily needs or for drugs that are not needed but which are addictive. Interestingly, there appear to be rostral-caudal dimensions in terms of approach and avoidance behaviors (Berridge, 2007), with diverse information signals playing more than one role in the organization of behavior-attention and learning.

Consider the anticipatory behaviors surrounding busy parents coming from work, juggling endless tasks as they return home to children, and parental experiences of endless humility, worry and ecstasy. Anticipation pervades this experience, like most of the cephalic adaptations that underlie our behavioral adaptations. Dopamine, one primary neurotransmitter amongst others, is tightly bound to diverse behaviors.

Diverse sets of neural systems are linked to the prediction of events and the recognition of uncertainty. Importantly diverse sets of neurons across dopaminergic pathways and neural sites are linked to either the expectation of rewards or the lack of predictable events. Two sets of dopamine have been tied to both events (Schultz, 2007). In other words, one set of dopaminergic cells tied to the basal ganglia have been linked to the prediction of reward-related events (e.g. dopamine; Schultz, 2007) and to incentive salience (Berridge, 2004).

Thus motor organization, pregnant with motivation, does not mean that they are the same, just as not all cognitive systems reduce to motor or vice versa, because motor systems are dependent on cognitive capacity; it just reveals the architectural design of the organization of action.

Again, steroid hormones such as cortisol, or estrogen-regulated dopamine, orchestrate motivation and attention towards affordable objects of relevance. Dopamine is broadly tied to the organization of action, inhibition, prediction of events knotted to the presence or absence of reward, feeding and sexual behavior, diverse forms of reasoning, non-reward but salient events, drug reinforcement, behavioral vulnerabilities (drug administration, anxiety, metabolism), associative learning and perhaps other forms of elementary learning, and just about any form of behavioral/cognitive task one can imagine. An example would be parental care, the organization of action and the expectations resulting in the release of dopamine in the central organization of action (Meaney, 2001).

Dopamine expansion, like other information molecules, is at the heart of our expansion in behavioral options; they offer greater options, more choice, and evolutionary advance, but also a chance for devolution. Too many choices, too many incentives, too many ways of making the wrong choice can, from a longer-term, well-being, viability point of view, be just as or more dangerous than too few. Just look at how it is tied anatomically across all the key areas and fundamental in the cephalic organization of action.

ANTICIPATORY REGULATION AND COGNITIVE/
VISCERAL ADAPTATION

As I have indicated throughout this book, anticipatory regulation underlies our evolutionally ascent and is a core feature of allostatic regulation. We are a species which is not simply reactive in the narrow sense of responding to a need or deficit after it happens; we anticipate the events. Regulation of the internal milieu has mostly been construed as restoring deficits, much less about anticipating them and the cognitive resources that reflect the cephalic innervation and maximal use of bodily resources towards long-term viability. But predictive responses are a primary adaptation in human evolution.

Homeostatic regulation is too passive a notion for the resources required to maintain long-term viability and reproductive success. Traditional homeostatics fail to consider brain involvement in shifting the emphasis towards longer-term considerations. The shift does not entail any mythological aggrandizement of perfect rationality, just the reflection of longer-term considerations in our adaptive responses.

The concept of heterostasis was introduced by Selye (1956, p. 85):

> Natural homeostatic mechanisms are usually sufficient to maintain a normal state of resistance. When faced with unusually heavy demands, however, ordinary homeostasis is not enough. The "thermostat of defense" must be raised to a higher level. For this process, I proposed the term heterostasis (from the Greek heteros = other) as the establishment of a new steady state by treatment with agents which stimulate the physiologic adaptive mechanisms through the development of normally dormant defensive tissue reactions. Both homeostasis and heterostasis, the milieu interior participates actively.

A number of other scholars have generated similar concepts.

- Moore-Ede (1986)
  - Predictive homeostasis is an anticipatory adaptation and is distinguished from "reactive homeostasis." This distinction arose in the context of considerations of circadian timing systems in the brain and their role in behavioral and physiological regulation in the anticipation of future needs when they appear.
- Mrosovsky (1990)
  - Coined the term "rheostasis," like the term "predictive homeostasis," it was generated to account for the variation in physiological systems, depending on season, time of day, and context.
- Bauman (2000, p. 85)
  - Used the term homeorhesis:
    The coordination of body tissue metabolism involves two types of regulation, homeostasis and homeorhesis. Homeostatic regulation involves the maintenance of physiological equilibrium; hence this control operates on a minute-by-minute basis to maintain constant conditions within the internal environment. Homeorhesis represents the orchestrated or coordinated changes in metabolism of body tissues to support a physiological state.

Allostatic considerations, like these others, shift the emphasis less in terms of maintaining a particular balance, for example ionic balance, and more in terms of the range of balances that are possible in an adaptive individual coping with a changing environment (Sterling and Eyer, 1988). Of course the concept of allostasis may not last the test of time in terms of utility. Homeostatic regulation is about maintaining the same conditions; allostasis emphasizes change. Change is the core feature of evolution and successful adaptation, and the ability to adapt to change is a core feature of our cephalic ascent. Fixed responses serve end points in environments that remain the same, and indeed a range of variation remains constant in long-term adaptation or viability.

Homeostatic regulation and its conceptual basis were originally bound to laboratory considerations (Cannon, 1916/1929). The depletion/repletion model became the standard way to tease apart the maintenance of the internal milieu (Richter, 1943). But once behavior was introduced as an essential role in sustaining long-term viability and not simple stability, a conceptual shift emerged. Behavioral adaptation and regulation is a fundamental way in which physiology viability is maintained; animals select food and mineral objects that are needed (Richter, 1943).

In fact, many investigators have argued that set point outlooks that are strictly homeostatic are not adequate to explain the behavioral role (Figure 6.3; Berridge, 2007). The hydraulic model of depletion and repletion is a limited perspective on the adaptive and opportunistic regulation of the internal milieu. It is not inaccurate, but there is greater variation in biological adaptation and long-term viability across different kinds of animals in nature than the homeostatic model allows for.

Moreover, the objects that a species require often shift toward hedonic reward. They are required, but they are also perceived as more palatable, thereby facilitating the motivation to acquire them. This is normal adaptation in the regulation of the internal milieu; in omnivores, like the rat, there can be over-consumption of salt-laden commodities under conditions of sodium hunger or under conditions of sodium repletion. There is no perfect homeostatic set point, there is just opportunistic ingestion, and certainly in our species, another omnivore, a vulnerability to over-consumption. Adaptation, in other words, is the norm, while homeostatic regulation is highly theoretical. One depiction of the classical homeostatic model is depicted below (Berridge, 2004). Information molecules that traverse both the peripheral and central nervous system underlie the organization of action, the recognition of the familiar and unfamiliar, the approach and avoidance of objects, and the evaluation of events. Neuropeptides are

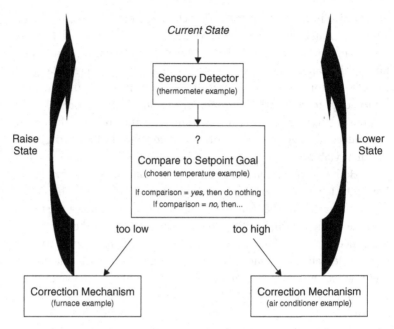

Figure 6.3   Hydraulic models of motivation. Homeostatic mechanism.
The mechanism uses negative feedback to correct errors in current
state. The current state of the moment is compared with the
set-point or desired goal value. If the current state is too high or too
low, an error-correcting mechanism is activated until the current
state returns toward the set-point. A room thermostat is an example
of a homeostatic mechanism based on set-point, goal comparison and
error detection, and negative feedback correction. In the thermostat
example, the set-point is the thermostat level you set, the current state
is the room temperature detected by the thermostat's thermometer,
and errors of room temperature are corrected by thermostat activation
of heating and cooling systems (reprinted from Berridge, K. C. (2004).
Motivation concepts in behavioral neuroscience. *Physiology and Behavior*,
81, 179–209. Copyright (2004), with permission from Elsevier).

organized around diverse behavioral functions, rarely only one; they
are too valuable a resource to be utterly restrictive to one function.

SALIENCE, OBJECTS, REWARDS AND DEVOLUTION OF
FUNCTION: GLUCOCORTICOIDS AND CRH

As our social milieu expanded, so did the range of objects for which
incentive motivation accrued. Regulation and anticipatory adaptation
have turned into vulnerability to exaggerated consumption.

Obesity, which is becoming a serious problem in the US and other countries, is an aberration of adaptive mechanisms gone amok: conservation and redistribution of food resources, attention to edible objects beleaguered by chronic flags of edible but ultimately unhealthy objects. Interestingly, the ingestion of saccharine can regulate CRH expression in rats; without corticosterone, the ingestion of the sweet taste directly moderates CRH gene expression in the PVN and the central nucleus of the amygdala (Dallman *et al.*, 2003) (Figure 6.4).

The hormones of energy homeostasis, of which there are many (e.g. leptin, NPY, CCK, insulin, ghrelin, corticosterone), contribute in diverse ways to energy viability. NPY can increase food consumption and leptin or CCK can decrease it (e.g. Smith, 1997). The hormones are not, for the most part, geared to excess since excess is a feature of modern times, and what occurs is devolution of function; one example of the vulnerability toward modern pathologies, such as obesity and adult onset or Type 2 diabetes. Diverse events impact these information molecules, evolved for adaptive energy conservation and viability; under some conditions, elevated levels of glucocorticoids following food deprivation or by glucocorticoid treatment can also increase NPY and food ingestion (Mercer *et al.*, 1996).

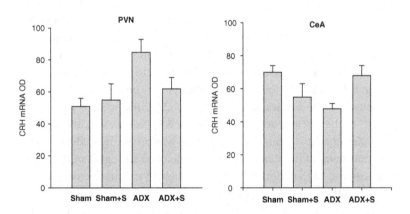

Figure 6.4   Adrenalectomized (ADX) rats have higher CRH concentrationin in the paraventricular nucleus (PVN) and lower in the central nucleus of the amygdala (CeA). When the rats consume a sucrose solution, CRH concentration becomes indistinguishable from that of control animals. (Dallman *et al.*, 2003).

Elevated corticosterone or cortisol is fundamental to foraging behaviors, to noticing events, via both the regulation of dopamine or norepinephrine or CRH, while it regulates the HPA axis and other end organ systems in both the peripheral and central systems. Of course, in an age of junk food abundance, metabolic pathology is symptomatic of meaningful interactions (Sterling, 2004).

The widening range of consumable objects turns the volume up in terms of arousing the physiology. The consumption of and withdrawal from known addictive substances, (heroin, cocaine, alcohol, barbiturates, amphetamine, marijuana, etc.), results in elevated levels of CRH in at least two regions of the brain: the central nucleus of the amygdala and the paraventricular nucleus of the hypothalamus. And perhaps glucocorticoid, at least with regard to amygdala function, may enhance CRH expression and thereby increase the vulnerability to increased salience and attention to those objects associated with these drugs of abuse (Schulkin, 2003). In fact, every object of reward (food, drugs, etc.) results in enhanced CRH expression in diverse regions of the brain, including the amygdala and frontal cortex and the paraventricular nucleus of the hypothalamus (Koob and LeMoal, 2005).

One way to understand these events is that one function of glucocorticoids is to magnify the effects of CRH in increased attention to objects, their potential value, etc. As I indicated earlier, one important insight into one behavioral function of CRH is that is released when an object is unfamiliar or uncertain, as well as if it is dangerous (Kalin et al., 1998). The neuropeptide is bound to attentional responses to both external and internal events. Peptides such as CRH can increase incentive salience (Pecina et al., 2006). Rats will bar press in anticipation of an expected sucrose reward following CRH infusion or amphetamine infusions into the nucleus accumbens. The next figure depicts responses to the intracranial infusion of CRH into what is called the "shell" of the nucleus accumbens (Figure 6.5; Pecina et al., 2006).

Recall that the nucleus accumbens was originally understood as the motor end of the limbic system. Gradually, this region of the brain has been found to underlie a number of behavioral functions essential for adaptation, including higher level cognitive functions that include probability judgments about statistical likelihoods, parsing of syntactic functions in language, and the organization of action and social responsiveness. The basal ganglia (of which the nucleus accumbens is a part) is a rather large region of the brain not simply motor, but it is not strictly one dimensionally cognitive; it is adaptive in many dimensions.

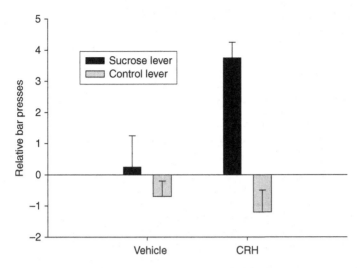

Figure 6.5   Injections of corticotrophin releasing hormone into the nucleus accumbens results in anticipatory bar pressing for sucrose (Pecina *et al.*, 2006, pubmed).

Many of these areas are activated with expectation of diverse rewards (e.g. sucrose rewards). For instance, expectancy for sucrose rewards results in activation of regions of the frontal cortex (Pecina *et al.*, 2006); in fact, both positive and negative events (good ingestion and drug withdrawal, for instance) activate CRH expression in diverse regions of the brain, including the amygdala, frontal cortex, and PVN (Koob and LeMoal, 2005). Interestingly, using microdialysis to measure CRH release corticosterone pretreatment potentiates the expression of CRH in both amygdala and frontal cortex (Figure 6.6).

Both consumption of and withdrawal from diverse psychotropic drugs elevates glucocorticoids and CRH (Koob and LeMoal, 2005); again, glucocorticoids like dopamine are essential for organization of action. In fact, that is one of the ways in which the steroid, by regulating the neurotransmitter, influences behavior. In this case, it is an all too real occurrence, namely the vulnerability to self-medication with drugs that are harmful. Feedforward systems include adrenal steroids and their regulation of CRH (or adrenergic action) in the attention and effort required that underlies the organization of action under adaptive and pathological conditions.

Drug abuse begins with reward, escalate to dependence and then to addiction, and finally to vulnerability to relapse. Reward declines and dependence expands; salience of cues grows as the importance

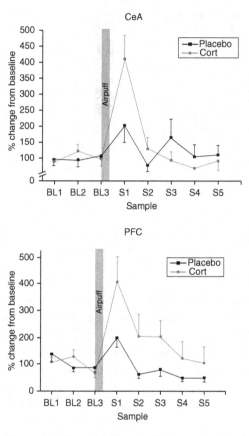

Figure 6.6   Interstitial levels of immunoreactive (ir)-CRH (expressed
as a percentage of baseline values) at the level of the central nucleus
and mPFC in different groups of animals under basal conditions and
following expoure to an airpuff stressor in vehicle (open symbols)
or corticosterone (solid symbols) pellet-implanted rats. A series
of consecutive brain sections bearing the trace of microdialysis
probes were stained, the most ventral location of the probe tip was
determined. Placements considered off-site were excluded from the
analysis (adapted from Merali *et al.*, 2008).

of the drug accrues and the broad based learning associating with drug use kicks in. Allostatic states reflect the escalation dependence and the decrease in reward. In other words, glucocorticoids regulate a number of neuropeptides and neurotransmitters bound to the organization of action. Importantly, glucocorticoids regulate a number of neurotransmitters, e.g. dopamine, bound to some aspect of reward or instrumental response. The regulation of dopamine, for instance, underlies the prediction of reward, the cognitive motor capacity underlying action; the regulation of norepinephrine underlies attention, the regulation of serotonin tonic mood and rhythm, all vital in the organization of action.

SOCIAL ADAPTATIONS, VULNERABILITIES AND ALLOSTATIC
OVERLOAD

The search for appetitive or social amelioration is satisfied by the consummatory phase of social attachment: social comfort. Adequate forms of social attachments in our species and in other primates are crucial to normal development, and many forms of social isolation lead to aberrant development. From the mundane to the extraordinary, hormones such as cortisol are knotted to the transition of mother or parental attachments (Ahnert et al., 2004) and in the maternal or parental response (Flemming et al., 1997). The events require picking up and retrieving; focusing attention requires energetic resources.

The non-adaptive result of dysregulated cortisol is the breakdown of end organ systems (brain, immune, fat cells, etc.), and thus a reflection of allostasis overloads (McEwen, 1998). The more chronic the events are, the greater the vulnerability for allostatic overload. Cortisol levels can go up in response to negative or positive events; the issue is to regulate the level of the hormone appropriately. These events have cephalic consequences. We know that dendritic expansion can occur in the amygdala and bed nucleus of the stria terminalis while they decline in the hippocampus (Mitra et al., 2005). These effects may be mediated by glucocorticoids and the induction of CRH and by TYPE 1-mediated CRH receptors; glucocorticoids can exacerbate neural deterioration (Sapolsky, 1990) and decrease in neurogenesis (Gould et al., 1996). Events in development can either create vulnerability or enhance neurogenesis in adulthood, creating a cephalic condition of less adaptive capabilities, a condition of allostatic overload.

Of course, there is also adaptation for forming friendships and alliances, perhaps reflecting a sex difference in a number of contexts.

The power to befriend is certainly a worthy adaptation for long-term viability and well-being. Friendship, as Aristotle and many others have noted, is essential for the human well-being, and tied to a number of health and well-being factors. Similarly, in an aging population possible only in a complex society such as ours, many of us are volunteering for diverse ways of giving back; we provide an "experience core" (Fried *et al.*, 2004), and giving away to others is a social model that also promotes well-being. Social bonds of meaning and comfort can also diminish PTSD (Charuvastra and Cloitre, 2008).

Social affiliation is one such long noted adaptation. It is not a mythologized panacea, but a piece of our biological/behavioral adaptations for which oxytocin is one information molecule likely to play diverse roles (Carter *et al.*, 1997/1999). A behavioral disposition towards more "dove"-like behavior rather than "hawk"-like behavior, to more collaborative and positive social cooperative interactions,may have real consequences (e.g. conduct disorder for the hawk and anxiety vulnerability for the dove on allostatic overload and vulnerability to physiological pathology (Korte *et al.*, 2004; Taylor, 2006). One type of allostatic overload has been thought to be linked to situations when energy demands capability; another when there is dysregulation of energy regulation linked to social functions (Wingfield, 2004).

Experiments with diverse rearing conditions (peer-raised animals versus those raised with surrogate or natural mothers) have shown long-term effects on behaviors and on diverse transmitter, peptide and steroid systems. A greater degree of CRH and lower levels of NPY are associated with greater fearfulness (Erickson *et al.*, 2005). A persistent elevation of central CRH is one feature that has emerged from diverse prenatal and postnatal adverse experiments (Meaney, 2001). In young macaques, for example, social uncertainty for resources (e.g. food) results in long-term elevation of central CRH; social contact and predictive features of resources including social contact reduce its expression.

One ameliorative vehicle for psychobiological well-being for our species is meaningful social contact; it underlies the regulation of diverse information molecules (e.g. CRH). The devolution of social contact negatively affects sleep patterns, cytokines, and cardiovascular and metabolic regulation (including Type II diabetes), and is not a prophylactic against diverse events that promote allostatic overload (Singer and Ryff, 1999). Long-term dysregulation of HPA and central CRH is lethal; the aging process is augmented by chronic adversity and elevated levels of cortisol, which degrade bone, brain and heart. It is not helped by diverse chronic social, economic health, danger and worries that permeate our social milieu, such as economic disparities; elevated cortisol is linked to social economic inequality, allostatic overload, and wear and tear on regulatory systems (Lupien, *et al.*, 2007). But cortisol should also be elevated in highly competitive individuals. The issue is whether the social milieu allows for appropriate secretion and non-secretion of this and other information molecules essential for adapting to diverse social contexts.

CONCLUSION

The price of an enlarged cortical and limbic system with greater motivational capacity is vulnerability to addiction. One of the original insights of Sterling and Eyer (1988) was the chronic overactivation of behavioral and physiological systems: lights on too long, too little sleep, too little enjoyment, too much seduction by noise, too little rest and restoration and meditation. Moreover, small satisfaction, quick fixes, and chronic wanting with little satiation or satisfaction, and endless desires associated with wanton junk, is no way to live or to maximize the utilization of adaptive mechanisms. Gluttony without end is a fall downward, a devolution of function.

In addition, context matters. The same stimulation of the brain can elicit different forms of behaviors depending upon the options in the environment, what the environmental affordances allow for (Gibson, 1979). One feature of our brain is the profound connectivity of forebrain with both the brainstem and the periphery. One synapse separates the gastrointestinal tract from forebrain sites. Massive connectivity of peripheral physiological end organ systems maximizes the utilization of anticipatory responses in the regulation of the internal milieu (Swanson, 2000). Information molecules such as peptide hormones are represented in the central and peripheral nervous system in addition to end organ systems, forming a continuous thread for regulation. Transmitters set the tone; neuropeptides are tied to function. Context, affordance, and opportunities are always a determining factor, along with physiological/cognitive capabilities.

Importantly, the diverse forms of information molecules will elicit different behaviors depending upon the context; CRH can elicit aggression or a fear-related response. But some of the information molecules are more encapsulated in function or narrow in function than others (e.g. angiotensin and thirst; Fitzsimons, 1979). In other words, the relative degree of specificity of function depends upon brain sites, social and ecological contexts, and the relative degree of specificity inherent in the information molecules.

What we know is that the expression of these diverse information molecules can easily turn from adaptation to pathology depending upon the context and the genetic vulnerability. Corticotropin releasing hormone, for instance, in extrahypothalamic sites where it acts as neuropeptide underling behavior, can turn from attention to endless distraction by objects to consume; a feature of devolution of functions is tied to addictive behaviors.

7

Neocortex, Amygdala, Prosocial Behaviors

INTRODUCTION

Diverse regions of the brain are essential for action and are often activated earlier than the action themselves; regions of the brain such as the basal ganglia (Swanson, 2000) are bound to diverse forms of cognitive actions, including prosocial ones. Diverse information molecules underlie prosocial behaviors (e.g. oxytocin, vasopressin, and serotonin).

A developmental perspective has emerged positing that neonates are rooted in the world of objects and transactions with others from birth (Kagan, 1984). Their aim is towards getting coherence in a social world; social cognitive dispositions predominate amongst other cognitive/physiological predilections essential for adaptation and coherence of action. Making sense of others is thus a core adaptation. So we come prepared to make sense of the objects around us – particularly con-specifics.

In this chapter, I discuss again corticalization of function, from a received view, not incorrect but perhaps overstated – namely that the cortex restrains the more primitive brain (e.g. amygdala). In addition, in this chapter, the role of diverse information molecules (oxytocin, CRH, dopamine, serotonin) in social approach and avoidance behaviors continues to be discussed, amidst a further understanding of both neocortical sites and amygdala function in the organization of action.

CONSIDERATIONS ABOUT THE NEOCORTEX

For over a century, we have known that cortical function is bound to more elaborate cognitive capacities; in fact, the nineteenth century was dominated by this realization. From a comparative perspective, the frontal cortex, for instance, makes transparent an expansion of the primate brain.

The frontal cortex is quite a large area of the brain. Of course many regions of the cortex (e.g. parietal cortex) have expanded. The roughly three pound organ sitting in a pound of fluid is an astonishing part of the body: metabolically voracious, a consumer of glucose and other vital energy-charged resources. For us, damage to diverse cortical regions promotes a devolution of various functions, including cognitive capacities that encompass the wide range of human choice, including memory and language. Flexibility of cognitive-behavioral functions and opportunity of choice in varying contexts is a key hallmark in our evolutionary cognitive ascent; damage to diverse regions of the brain, not just the cortex, dilutes this capability. Broad-based anatomical connectivity between orbital and prefrontal cortices is prominent, along with the amygdala and basal ganglia, to mention several regions of importance (Swanson, 2000). These regions of the cortex and their input and output to vital limbic/motor areas underlie an amazing array of cognitive/behavioral functions.

The tasks in which the neocortex is involved are as diverse as we are as a species. In fact, the list is exhaustive of primate cognitive activities and our species in particular, as we have become more dependent upon cortical functions; this has been known for nearly two centuries. We have grown more sophisticated; we monitor outcome measures by linking regions of the cortex and tying into basal ganglia for the evaluation of incentives and outcomes (Berridge, 2007; Schultz, 2007). It gets more manageable and interesting when traditional areas such as Broca's area are linked to broader classes of syntactical events, such as musical composition and underlying expectations of form and function within musical composition (Muess *et al.*, 2001).

Of course, the story is not that simple, since both regions will be linked to both sorts of functions. But the suggestion is interesting; thus, for instance, using brain imaging methodologies (PET, fMRI), regions of the cingulate cortex are shown to underlie the orchestration of conflict and uncertainty (Dolan, 2007). In PET studies, there is greater blood flow to this region during conflict and adaptation. Monitoring behavioral selection parameters results in the organization and effort that underlies the realization of our plans and projects. The events are bound to uncertainty about expectations, all of which are tied to the cingulate cortex.

In electrophysiological studies in macaques, choice of action is linked to the anterior cingulate cortex (Isomura *et al.*, 2003). The cingulate cortical regions have long been linked to limbic functions (Damasio, 1996). In addition to brainstem visceral projections to this region, there is massive neural connectivity between cingulated cortex and diverse cortical regions, including the frontal cortex (Swanson, 2000). Both frontal and cingulate cortexes have been linked to reward expectancies (Rizzolatti and Luppino, 2001).

However, many regions of the brain are activated under conditions of conflict and discrepancy of expected events including regions of the cingulate cortex (Schultz, 2007). The greater the difficulty, the greater the conflict; the greater the discrepancy, the greater the activation of neural response. The amygdala is also activated under these conditions.

THE VIEW

One common view, not necessarily mistaken, but perhaps a bit misleading, is that primitive behaviors emerge as the cortex devolves. With less cognitive capacity, there is less neocortical control and more limbic expression. This has been the dominant view from the nineteenth century to the present. Neural systems for control are neocortical; the impulsive and less reflective responses require fewer cognitive resources and are subcortical (Jackson, 1884/1958). This again is the standard neuroscientific view, not wrong, but overstated and under-tested.

Altering the received view are modulating influences from diverse neurochemical systems (e.g. serotonin, dopamine). Serotonin is an ancient molecule found in both vertebrates and invertebrates (Strand, 1999) which provides a modulating regulator signal throughout the nervous system, including the neocortex, and that underlies the adaptation to choice and conflict.

A frequently studied temperament style is that of behaviorally inhibited children. Such children demonstrate increased vigilance and uncertainty as well as heightened reactivity to novelty, which are accompanied by an increased amygdalar response (Schwartz *et al.*, 2003). Others (Schmidt *et al.*, 1999) have found that behaviorally inhibited children have more activity in the right than the left frontal cortex, an effect that has been found in nonhuman primates as well. People with greater right frontal activity exhibit an increased reactivity to negative stimuli, demonstrated as behavioral inhibition and vigilant attention – a social withdrawal response. Greater left frontal activity is associated with greater positive affect and greater behavioral activation and goal-approach behaviors – an approach response (Davidson *et al.* 2003).

Again, shy, fearful children in a social context show greater right preferential activation to social scenes (Davidson *et al.*, 1997), and higher levels of cortisol (Kagan *et al.*, 1988) following a social context. This is expressed very early on in ontogeny within a few months postnatal (Schmidt *et al.*, 2007). These relationships have also been demonstrated in macaques (Kalin *et al.*, 1998). Serotonin expression and regulation may be linked to shyness (Schmidt *et al.*, 2007). Changes in the serotonin gene structure (long and short version of the 5-HT receptor and dopamine) region have been suggested to be linked to temperamental shyness and behavioral inhibition (though the link to behavior does not account for much of the variance) and altered frontal neocortical lateralization of function (Figure 7.1).

Importantly, regions of the brain rich in information molecules are tied to social assessment. In the instance of unfairness, manipulations of 5-HT function influence the sense of unfairness; lower levels are reported to increase retaliation towards others; pharmacological depletion of serotonin increases retaliatory responses to perceived unfairness. The investigators suggest that these effects reflect ventral prefrontal cortex damage (Crockett *et al.*, 2008).

Of course serotonin, like other broad neurotransmitters, underlies diverse behavioral adaptations, and deviations of normal gene

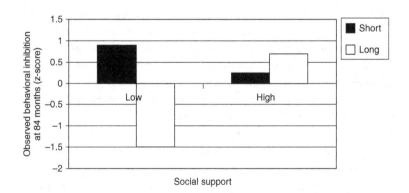

Figure 7.1   Some evidence for a gene-environment interaction in predicting behavioral inhibition in children (long and short forms of serotonin receptors (Schmidt *et al.*, unpublished).

function can tip the balance towards devolution depending upon the social context: the tone of a response as serotonin, the attentional requirements as in norepinephrine, and dopamine for response to incentives and the organization of action. Interestingly, serotonin transporter variation is linked to amygdala function and the regulation of social fear, and is perhaps disinhibited by neocortical devolution of function (Hariri *et al.*, 2002).

THE CORTEX AND THE ADJUDICATION OF CONFLICT

There are diverse cortical (and perhaps not simply neocortical) sites participating in the adjudication of these diverse competing interests and motivational pull and incentives' lures. Thus, regions of the neocortex are activated and compete for expression during choice and conflict. Two regions of the neocortex stand out in the context of conflict adjudication and the organization of decision making – the cingulate cortex, vital in adjudication of competing interests, and the frontal cortex in its implementation (Dolan, 2007)

Control and cortex have become somewhat synonymous, particularly with the frontal cortex (Jackson, 1884/1958), and for this there are diverse forms of evidence that date back several years now (Botvinick *et al.*, 2001). Of course inhibition and cortical control does provide opportunities, one of which is to restrain behavior; this is the core cortical perspective in light of evolution and the affordance of diverse behavioral options (Jackson, 1884/1958; Broca, 1878), including essential bodily ways in which to engage and integrate information. More generally, fast driving visceral representations are fundamental and grounded in cognitive systems and the organization of action.

MORAL CONFLICT, CONTROL, CORTEX

Diverse forms of moral considerations in which personal and nonpersonal configure are commonplace. An extreme laboratory experiment makes this plain. The trolley car experiment is an example of personal conflict: having to push as opposed to ordering someone from afar is all the difference in the world. The result is greater conflict because the event is up close and personal (e.g. should I opt for killing one innocent to save five other lives and suffer the angst of being a murderer forever, or should I abstain from doing so and therefore regret being responsible for the death of five people because of my own omission?).

Several regions of the neocortex were activated in the more conflicted context. Indeed, the more conflicted the situation, requiring cognitive processing (e.g. utilitarian decision-making where one has to summate possibilities), the greater the cognitive load on prosocial reasoning, and the more it may require frontal cortex.

Damage to the frontal cortex has been linked to a devolution of moral sentiments and moral judgments, sociopathic behavior and a devolution of empathy towards others (Damasio *et al.*, 1990). Indeed we now know moral judgment activates a broad array of brain regions that closely overlap, not surprisingly with regions knotted to prosocial behaviors. But in fact, the PFC is engaged *whether or not* decisions or behavioral outputs are required in moral scenarios, suggesting that the PFC does not merely manipulate information stored elsewhere in the brain, but in fact represents certain aspects of social knowledge or action.

In addition, patients with ventromedial prefrontal cortex (vmPFC) damage made more utilitarian choices in trolley-type dilemmas; they opted more often for costly punishing non-cooperators in the ultimatum game, i.e., they were more emotional. One possibility is that an intact vmPFC is more critical for experience of prosocial sentiments associated with affiliative components (i.e. guilt, compassion, interpersonal attachment), whereas the dorsal lateral prefrontal cortex (dlPFC) and lateral orbital frontal cortex (OFC) are more relevant for other-critical sentiments (such as indignation and contempt) (Moll and Schulkin, 2009). In other words, reduced prosocial sentiments are compatible with increased utilitarian choices in personal dilemmas, and preserved or increased punishment of others in the ultimatum game (Koenigs *et al.*, 2007). Indeed, the greater the cognitive conflict or cognitive load (Greene *et al.*, 2008), the greater the information to be processed, and the greater the activation of regions of the frontal cortex (dlPFC and anterior PFC and certain posterior parietal regions) (Greene *et al.*, 2004). But wide arrays of brain areas are tied to moral judgment that spread over diverse regions of the neocortex in addition to subcortical regions that underlie more generally prosocial behaviors (Figure 7.2).

CORTICAL INHIBITION

The prefrontal cortex (PFC) has been implicated in the inhibition of fear responses (Morgan and LeDoux, 1995), and is also associated with a number of disorders: anxiety disorders, such as PTSD, and substance use disorders. Some of the more consistent characteristics of frontal lobe damage include decreased self-monitoring and disinhibition or impaired impulse control (Bechara, 2005).

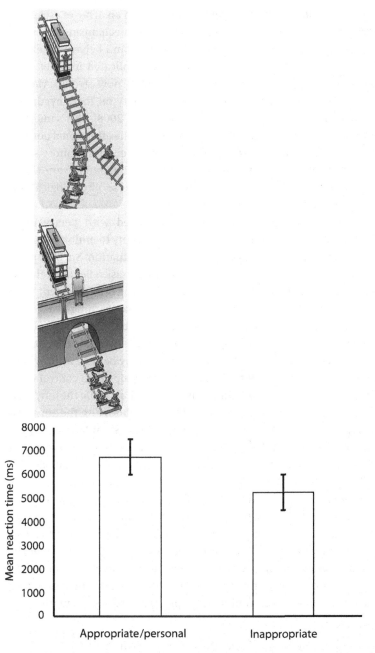

Figure 7.2   The greater the conflict, and perhaps more things to consider and more personal, the longer the reaction and the greater the activity (from Greene, J.D., Sommerville, R. B., Nystrom, L. E., Darley, J. M., Cohen, J. D. (2001). An fMRI investigation of emotional engagement in moral judgment. *Science*, **293**, 2105–8. Reprinted with permission from AAAS).

Inhibition is a rich concept, and is used in an array of areas; it is associated with reversal learning, attentional mechanisms, the study of temperament in shy children, extraversion, and behavioral extinction. The prefrontal cortex (PFC) has been implicated in all of these examples of inhibitory function, and there is now strong evidence that PFC may be exerting its inhibitory capacity on the amygdala in the case of extinction learning (Delgado *et al.*, 2008); this same PFC-amygdala interaction may also be occurring in the attentional domain. One of the mechanisms is via GABA, a fundamental inhibitory signal on end organ tissue including brain; GABA and glutamate are widely distributed across the brain and underlie diverse physiological and behavioral functions (Quirk *et al.*, 2000).

PFC pathology has long been associated with perseverative responding, which may be described as an inability to inhibit behaviors that are no longer appropriate in the current situation. Such responding was typically observed in reversal learning tasks. Behavioral flexibility, or the ability to adjust responses when the reinforcement value of a stimulus changes, is reflected in the reversal learning task.

It frequently has been suggested that brain regions such as the PFC are involved in regulating amygdala generated emotional responses (Quirk *et al.*, 2000). The amygdala may be seen as providing rapid threat detection, with the PFC providing top-down control when given sufficient processing time. It is thought that, while the amygdala forms initial fear associations, the mPFC is inhibitory on the amygdala during extinction (Morgan and LeDoux, 1995). Stimulation of mPFC afferents to the lateral amygdala (LA) in rodents suppressed LA activity evoked by conditioned stimuli, leading the authors to conclude that PFC may regulate affect through inhibition of the amygdala.

Specific to extinction, it was found that lesions of the auditory and visual (LeDoux, 1995) cortex left acquisition intact while greatly prolonging conditioned responding during extinction. One reasonable suggestion is that subcortical input to the amygdala may mediate learning which is relatively permanent, while the cortex modulates expression of these permanent memories; lesions of the cortex thus may have allowed the uninhibited expression of these memories

DOPAMINE AND THE ORGANIZATION OF INHIBITION: MATURATION OF FRONTAL CORTEX

Prefrontal dopaminergic and immunoreactive dopaminergic neurons are fundamental to memory tasks in the organization of action;

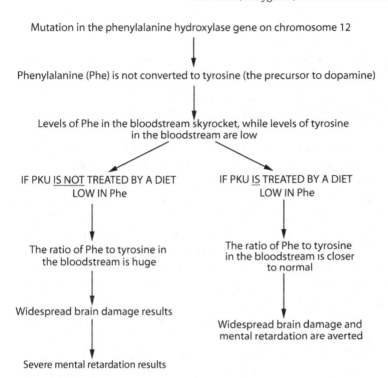

Mutation in the phenylalanine hydroxylase gene on chromosome 12

Phenylalanine (Phe) is not converted to tyrosine (the precursor to dopamine)

Levels of Phe in the bloodstream skyrocket, while levels of tyrosine in the bloodstream are low

IF PKU IS NOT TREATED BY A DIET LOW IN Phe

IF PKU IS TREATED BY A DIET LOW IN Phe

The ratio of Phe to tyrosine in the bloodstream is huge

The ratio of Phe to tyrosine in the bloodstream is closer to normal

Widespread brain damage results

Widespread brain damage and mental retardation are averted

Severe mental retardation results

cognitive emboldening in movement is an important adaptation; dopaminergic control from cortical regions is tied to behavioral inhibition (Goldman-Rakic et al., 1989).

Our evolutionary legacy is bound to diverse forms of inhibition and adjudication of competing drives. The broadcasting of many different competing interests permeates central states resulting in the inhibition of some interests and the expression of others. Dopamine expression is essential in the development of inhibition (and more generally the organization of action) in our species; aberrations in dopamine expression undermine this important cognitive/motor capacity. For instance, children born with PKU syndrome, which degenerates a conversion dopamine, often have impairments in behavioral inhibition essential for normal development. Experimentally, this has resulted in delayed alternation, or delayed matching task, in the laboratory. This is a task that requires inhibition and is bound to the development of PFC and dopamine expression. There is a developmental event in which this is easily expressed by children by maturation in the brain (Diamond, 2001). The cognitive control over behavioral inhibition is also impaired with damage to PFC in macaques (Diamond, 2001).

THE AMYGDALA

We turn to the amygdala again now. The amygdala is phylogenetically ancient, and expressed in all vertebrates. It is under the temporal lobe and is almond-shaped by tradition; but the boundaries of the region are highly conceptual and therefore can be subdivided in a number of ways. The amygdala plays a role in diverse adaptive behaviors essential for social function, including approach and avoidance behaviors and appetitive and exploratory behaviors (Aggleton, 1992/2000).

Recall that the more medial and central medial regions of the amygdala are tied to olfactory and gustatory information processing; the more lateral regions are linked to more associative and integrative functions (Norgren, 1995). Thus, the size of the lateral region of the amygdala, and not just neocortical regions, reflects the range of functions for which it plays a role (LeDoux, 1995). The amygdala is a diverse, heterogeneous region, and can be defined a number of ways (Swanson, 2000). Depending on how one understands the amygdala, it contains from 5–7 nuclei. In evolutionary terms, all vertebrates have a clear and distinct amygdala.

While there may be disputes with regard to what the amygdala really encompasses structurally, functionally, it is a classical limbic region with diverse regulatory functions that cover a wide span, from mineral ingestion, to fear, sex, and recognition of social relations. For Nauta (1972), it was the link between motivation and motor action through projections to the nucleus accumbens region. And of course there are massive projections from the amygdala to the bed nucleus of the stria terminalis and from regions of the bed nucleus of the stria terminalis (e.g. central and medial nucleus) to these same regions within the amygdala.

The amygdala is more labile than generally depicted. Just as emotional assessments are not fixed, there is flexibility in amygdala responses to diverse events. Flexibility is a feature of adaptive responses in changing social networks and contexts; devolution of function is a feature of pathology (Jackson, 1884/1958), less social contact, and less flexibility in modulating a more limited range of options. The traditional view of the amygdala is that it is part of the olfactory and limbic cortex. All olfactory signals reach the amygdala, but then again most sensory systems reach the amygdala and, in particular, gustatory, visual, and cardiovascular systems (Norgren, 1995). It is a major computational sensory relay (Aggleton, 1992/2000), and we know that many of the neuropeptides that underlie diverse behavioral adaptations are produced in nuclei within the amygdala. Mapped out in great detail by the classical anatomist C. J. Herrick in the brain of catfish, the amygdala was understood to be greatly involved in visceral regulation, taste and olfaction. It was originally thought of as crucial to smell, then later a vital part of the visceral nervous system of the limbic system.

TASTE AND DISTASTE IN SOCIAL CONTACT

Gustatory visceral disgust is primordial and related to moral revulsion, in addition to other forms of revulsion (Rozin, 1998); it is a primary cephalic vehicle, easily conditioned to events and to forms of contagion by association to food sources that have rendered one ill (Rozin, 1976) and expanded perhaps to include moral revulsion, as a strong visceral signal. It is an important preadaptive mechanism.

Disgust evolved and is designed for one set of tasks – avoidance of foul foods – and is a fundamental link to cephalic systems underlying visceral distress. It has been extended to new and varied domains, such as the avoidance of people (moral disgust). Disgust is linked to condemnation or moral revulsion, and many regions of the neocortex (e.g. medial and orbital frontal cortex) and other regions of the brain (e.g. striatum, amygdala) are activated by it.

Taste is a primary projection to the central nucleus, and in fact all sensory systems traverse amygdala function (Norgren, 1995). One pathway of visceral/taste projections reaches the central nucleus while another distinct pathway through the thalamus reaches neocortical tissue (see also Chapter 3; Norgren, 1995). One pathway may be more involved in gustatory/visceral discrimination (thalamic and gustatory cortex), the ventral pathway with the more motivational (amygdala) features that underlie approach/avoidance decisions (Norgren, 1995). It is interesting that several fMRI studies find basal ganglia activation to disgust reactions (Calder et al., 2000) in addition to gustatory cortex when shown disgust faces (Phillips et al., 1998).

Importantly, cephalic innervation into the periphery is a critical evolutionary piece in the anticipation of environmental events and the regulation of the social milieu. The connectivity between regions of the frontal cortex and amygdala is intense, and there is also connectivity between the regions of the amygdala and peripheral sites.

Visceral representations are an example of critical information about the gastrointestinal tract. It is one neuron away from both the solitary tract of the brainstem and the gut itself (Norgren,

1995). Regions of the solitary nucleus contain site specific representations of end organ systems of the alimentary canal. Direct connectivity of the solitary nucleus to regions of the frontal cortex and amygdala places these regions within one or two neurons from the alimentary end organ systems. This is a critical adaptation when the complexity of the social milieu requires a broad array of information processing systems in which "gut reactions" (i.e. getting the "gist" of something) are a critical piece of the cognitive/physiological assessment.

AMYGDALA SOCIAL FUNCTION

Damage to the amygdala region compromises the perception of rudimentary actions; the perception and action of linking the site of an object with an appropriate species specific response is one classic example –the "Kluver Bucy Syndrome." In fact, damage to this region disrupts diverse forms of behavior, including social behavior (Kluver and Bucy, 1939). More recently, a set of studies by Emery and Amaral (2000) have demonstrated that this region of the brain is essential for normal social behavior. An emphasis on impairments in social perception has long been understood as entailing temporal lobe damage, of which the amygdala is often construed as part. In other words, amygdala damage compromises diverse social functions (Kluver and Bucy, 1939) and for perhaps getting the "gist" of an event, for making contact with others, and the assessment of the value of an event.

Of course, Darwin emphasized taxonomic depictions of facial response as broadly understood biological features that underlined and were pivotal for social activity in our species. Since Darwin, it has been noted that an important source of information for our species is facial expression; eye contact in particular is an important root of fundamental social contact (Kagan, 2002). Facial expressions generate approach and avoidance behaviors inconsistent with the value of the context; of course this is consistent with the classical view of this region being tied to the evaluation of events. For example, damage to this region impairs the ability to detect safety and approachability; in a subject with complete amygdala damage the poor use of eye information from the region of faces was notable. However, when given formal instruction to focus on the eyes, this defect was corrected (Adolphs *et al.*, 2005)

Importantly studies show that amygdala damage in humans decreases eye contact in conversational context, which is also a feature of autism (Baron-Cohen, 1995/2000).

One feature of amygdala function is its link to fear and the unfamiliar; amygdala damage results in impaired responses to fear-related stimuli and facial expressions (Adolphs, 1999). A window into others, into the social milieu, eye contact is impaired in these individuals. Therefore, they perhaps experience more fear because of less social contact and alteration of fear circuitry. A hyperexcitable amygdala is linked to diverse forms of fear-related behaviors, and contributes to social anxiety (Kagan et al., 1988) and heightened responses to fearful and uncertain or unfamiliar faces (Schwartz et al., 2003).

Social learning is diverse, and one important feature is what to avoid and what to fear. Some of these social fears or phobias are built into cephalic organization through prepared kinds of fast learning, some of which are linked to facial expressions (Figure 7.3; Morris et al., 1996).

Fast and inappropriate prejudicial responses are also linked to amygdala activation. A broad-based behavioral dimension is the familiar and unfamiliar distinction (Kagan, 2002). The unfamiliar provokes attention and learning and scientific inquiry. The amygdala is activated under all these conditions, and amygdala activation is linked to a broad array of emotions (Phillips et al., 1998). More generally, the amygdala is part of a neural circuit for assessing events, for approach and avoidance of affordable or useful objects, and linked to play-related behaviors in development. Play behavior is a conduit towards others, for learning from others, and the development of species-specific social skills. As might be expected, adults characterized as inhibited as children had increased amygdala reactivity to novel events (Schwartz et al., 2003). Interestingly, while shy/inhibited children showed greater activation to a novel event, bold children demonstrated more activity in the accumbens to a novel event (Beaton et al., 2008).

It is known from diverse fMRI studies that the amygdala is activated to something unfamiliar, and to something that has a social dimension (Beaton et al., 2008). Cortisol infusions influence diverse forms of memory by activating the amygdala, and neurotransmitters, such as norepinephrine (or CRH [Schulkin et al., 2005]), in diverse brain regions influence memory; indeed, cortisol infusions impact amygdala activation and the response to diverse facial responses, in particular sad faces (Erickson *et al.*, pers comm.). The lateral region of the amygdala, which is a more evolved region that underlies most forms of fear-related behaviors (LeDoux, 1995), and for which receptors for diverse information (e.g. glucocorticoid, CRH) exist, are known to interfere with normal processing when antagonists are infused centrally (McGaugh, 2000).

However, the dimensions of unfamiliar and familiar are broad-based behavioral categories, like approach and avoidance, that map more generally in the organization of action (Kagan, 2002). It is not surprising that the amygdala is activated by unfamiliar faces and fast judgments, and the more complicated the facial expression the greater the activation of the amygdala in addition to other regions (Beaton *et al.*, 2008).

SOCIAL JUDGMENT, OXYTOCIN AND THE AMYGDALA

Fast social judgment underlies human decision-making and is linked to amygdala function. But social attachments are footholds by which we connect to the larger world and get anchored towards others – a social network of genes anchored towards social attachment, which include oxytocin and vasopressin underlying part of the social brain. Diverse genes such as oxytocin contribute towards this end.

Regions of the brain long linked to reward and social perception (e.g. the amygdala) are bound to social judgment more generally and for getting the "gist of things" (Adolphs, 2001) or quick heuristic judgments. Such fast judgments underlie many social decisions; diverse forms of cognitive adaptations emanate from a prepared cephalic system. Underlying a diverse set of cognitive adaptations are rich sensory/motor systems pregnant with the ability for getting the gist of things. Some of these adaptations are tied to amygdala functions.

Moreover, the amygdala, amongst other regions, underlies the perception of others' intentions, beliefs and desires. For instance, face perception and eye contact are linked to amygdala function (Rolls, 2000). This ability to understand the beliefs and desires of others is fundamental to our social knowledge, our getting anchored to the social world of others. For us, social contact is made in part through visual representation, so it is not surprising that oxytocin is linked to shared functions and social attachment (Hurley, 2008). Both social support and oxytocin administered intranasally decreases levels of cortisol in normal human subjects, exposed to a social stress. Social contact, including physical closeness eye contact, social listening, etc.

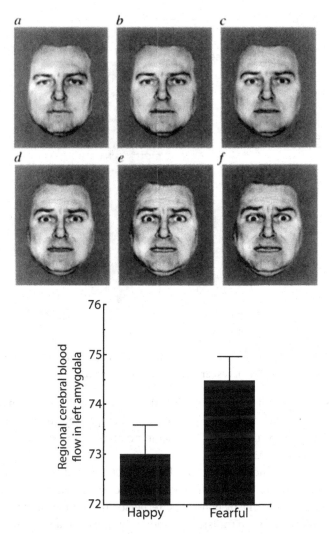

Figure 7.3   Prototypical neutral and fearful expressions and amygdala activation (adapted from Morris *et al.*, 1996).

provides an important part of our well-being, and the social regulation of each other. Oxytocin is one information molecule tied to social contact.

In humans, for instance, oxytocin inhalation facilitates social recognition and social memory (Savaskan *et al.*, 2008). Importantly, higher oxytocin levels are associated with increased likelihood of reciprocation in a game in which trust is quantified. The higher the levels of oxytocin, the greater the trust in an economic game; oxytocin

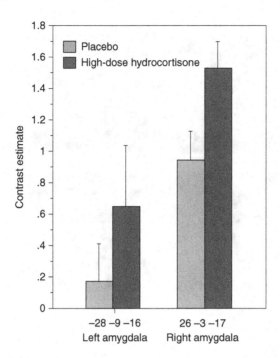

Figure 7.4   Differential response to cortisol in the amygdala in subjects when shown sad faces (Erickson *et al.*, pers. comm.).

levels were found to be higher in subjects who received a monetary transfer signaling an intention to trust, in comparison to an unintentional monetary transfer of the same amount from another player (Kosfeld *et al.*, 2005).

A study conducted by Kosfeld *et al.* (2005) found that intranasal administration of oxytocin induced more cooperation in an anonymous economic game by boosting interpersonal trust. In this game, the first player chooses to transfer an amount of money (if any) to another player. The amount is multiplied, and the second player may choose how much he/she will transfer back to the first player (i.e. reciprocation). Exogenous oxytocin administration was associated with increased amounts of transfers in the trust game by first movers.

Intranasal administered oxytocin also enhances the memory of faces versus mechanical objects; oxytocin facilitates the memory of faces (Rimmele *et al.*, 2009). Social support and increasing oxytocin in human studies have been shown to decrease trauma in (PTSD) patients (Heinrichs *et al.*, 2003). Decreasing social anxiety or fear might also be an important effect of oxytocin, a hypothesis that was strengthened

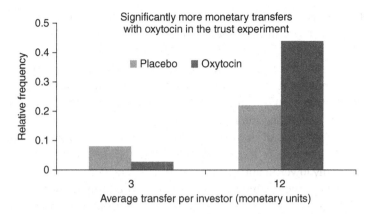

Figure 7.5   Significant decreases in coupling of the amygdala to the midbrain under oxytocin. Plot of oxytocin effect on correlation with the amygdala at the midbrain location showing maximum linkage to the amygdala during the placebo condition, highly significant decrease under oxytocin (adapted from Kirsch *et al.*, 2005).

by a recent pharmacological fMRI study. In this study, Kirsh and colleagues showed that oxytocin decreased amygdala activation to fearful stimuli (Figure 7.5; Kirsch *et al.*, 2005).

Other studies have shown that oxytocin impacts the affective value of conditional fear associated with events (Petrovic *et al.*, 2008). Importantly, direct eye gaze (for instance, what can be threatening) results in amygdala activation; it is decreased by oxytocin given trannasally in humans in whom fMRI was used to measure brain activation (Petrovic *et al.*, 2008). More generally, oxytocin reduces the impact of emotional faces on amygdala activation (Singer *et al.*, 2008) and facilitates recognition of words about relationships (Unkelbach *et al.*, 2008).

DEVOLUTION OF SOCIAL FUNCTION AND AUTISM

Autism is, amongst other things, a loss of social contact and is linked to amygdala function. Basic social cognitive systems, such as joint contact, are compromised in autistic individuals (Baron-Cohen, 1995/2000). Studies have shown abnormalities in diverse social contexts requiring normal amygdala function. One feature of autism is the devolution of function with regard to the recognition of different mental states (Baron-Cohen, 1995/2000), even when controlling for intelligence functions between normal and autistic individuals. Without this function, an important system is compromised: a fast heuristic.

Of course, diverse regions of the brain are linked to the devolution of social function noted in autism, and they include the frontal and temporal cortex and cerebellum; the fusiform cortex is linked to facial processing (Beaton *et al.*, 2008), and perhaps fear is linked to devolution of function in autism as well.

Decreases in oxytocin have been noted in autistic individuals, which may reflect less social contact. Indeed, levels of oxytocin are generally lower in autistic individuals (Carter, 2007).

Importantly, infusions of oxytocin have been found to ameliorate some of the abnormalities associated with autism. For instance, autism is associated with repetitive behaviors. Oxytocin reduced the repetitive behaviors and also enhanced social contact and cognition (Hollander *et al.*, 2003).

CONCLUSION

One feature of corticalization of function and our evolution is greater expressive value, and more cognitive resources to draw on; and with anatomical accessibility of cortical input to diverse regions of the brain, an expansion of behavioral options became more available (Rozin, 1976). A distributed system, and not strictly a hierarchical one, appears to be part of a conception of the wiring diagram of the brain (Berridge, 2004; Bernston and Cacioppo, 2000). A distributed neural system with vision being prominent, and visceral connectivity from the brainstem to cortex (both old and new), is a cardinal feature of the neural architecture.

The cortical role of inhibition is a feature of our evolution, a view held from the nineteenth century to the present; but the neocortex is both inhibitory and excitatory, as are other regions of the brain such as the amygdala. Critical information molecules, like dopamine, underlie behavioral inhibition critical for development and self regulation, but they also underlie the organization of action and the prediction of events.

I have highlighted two regions and several transmitters and neuropeptides, but the list is longer and a full understanding will include the wide range of brain regions and information molecules (e.g. dopamine, serotonin, dopamine and oxytocin). Critically, neuropeptides like oxytocin tend to decrease fear related events, and also decrease amygdala activation to fear-related stimuli; perhaps it is less about fear and more about an acceptance of the need to explore unfamiliar events, to decrease social wariness.

Devolution is the converse of evolution of the central nervous system (Jackson, 1884/1958); behavioral functions decrease in scope, expression, and competence, something that is apparent in the devolution in cephalic capacity and the onset of autism.

Conclusion: Evolution, Social Allostasis and Well-Being

INTRODUCTION

We have come to realize, as a normative goal, that social hope is our common bond – the dilution of differences that divide and harm us. The path of human progress is frail; with glimmers of hope, the eternal seduction, the stoic nobility exists amidst the diffidence and difficulties in preserving a broad social compass in which many reach forms of meaningful human happiness rich in existential sensibility.

However, there is no panacea; the idea of progress that infused Darwin's conception of evolution and Jackson's and Spencer's conception of the nervous system has been modified. Corticalization of function does not necessarily mean social advance. Devolution of function is as paramount as times of war and crisis.

Prosocial sensibilities figure importantly and are a constant across cultures, but they compete with diverse motivations. Variation in expression is a constant, but social contact is also a factor across all cultures, and the formation of habits sets the conditions for meaningful lasting social contact (Jaspers, 1913/1997).

EVOLUTION AND ADAPTATION

At one fundamental level, little has changed: we search for the stable amidst the precarious (Dewey, 1925/1989). The search requires diverse cephalic and cultural resources, and results in punctuated and gradual cultural epicenters. The human condition remains more precarious, our weapons that much more dangerous, and the level of potential destruction that much greater. The precarious shifts towards the more stable by cephalic adaptation. Core needs are always a common function satisfied by food, water, sensual contact, sport, explorations, etc; the diverse motivations that underlie these needs are quite broad, as we are broad in potential for expression (Carter *et al.*, 1997/1999). Primates are small in number but powerful in their impact on the earth.

What evolved in our species are long-term social bonds, plasticity of expression, and corticalization of function (Figure C.1). And as our cortical visual functions increased dramatically, standing up and looking and forming eye contact began an evolutionary expansion in many primates. Human social contact, representation of objects, and use of objects are core cognitive capacities; technology is an extension of ourselves, expanding what we explore.

In addition, regions of the amygdala essential for social attachment and avoidance also demonstrate significant changes in us: for instance, enlargement of the lateral amygdala, which is closely tied to neocortical function (Swanson, 2000). The largest nuclear region is the basal lateral region. In one comparative study of apes and humans (e.g. human, chimpanzee, bonobo, gorilla, orangutan, gibbon), investigators found that the size of the lateral division of the amygdala expands quite a bit in homo sapiens compared to the expansion in other primates (Figure C.2; Barger *et al.*, 2007).

Innovative tool use is not surprisingly linked in diverse primates to an expanded neocortex.We come prepared with an arsenal of cognitive adaptations rooted in social discourse and commerce with one another and the construction of objects that we use – our tools. And our evolution is knotted to social groups working in unison across diverse terrains. Key abilities include discerning the wants and the

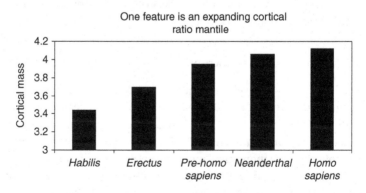

Figure C.1   Expanding cortex (adapted from Foley, 1995).

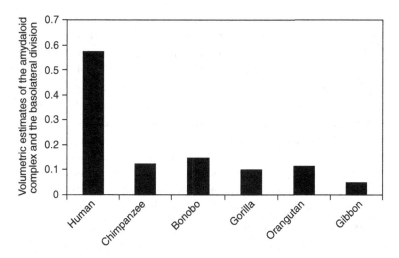

Figure C.2   Volumetric estimates of the amygdaloid complex and the basolateral division (adapted from Barger *et al.*, 2007).

desires of others (a core feature of our adaptations), along with cognitive adaptations such as recognizing the kinds of objects that are useful or affordable and avoidable, coupled with a wide array of inhibitory capacities that contribute to social cooperative behaviors.

Cephalic expansion set the stage for technological creations, expanding our sensory systems. Seeing by magnifying became an evolving theme as our capacities were extended and we turned from managing nature towards understanding nature – tool use and tool making was critical for this development.

HUMAN EMPATHY, DEVOLUTION OF EMPATHY AND
SOCIAL CONTACT

Empathy is just one of the moral sentiments, but is an important cognitive adaptation and a cardinal feature of our evolution. A social brain is distributed across a wide array of neural structure and function devoted to negotiating complex social interactions.

A conception of the experiences of others is one frail but important psychobiological predilection to reduce morally offensive behaviors. Experiments in the laboratory and real life point this out. In experiments where physical contact is required, it is much more difficult to sacrifice one for the many (Greene *et al.*, 2004). The more personal the social contact, conversely, the more conflict that can emerge.

Cognitive dissonance is one mechanism to reduce conflict by segmentation and rationale, justified or not (Sabini and Silver, 1982).

But "empathy" only goes so far; thus, one cannot rest too much on that one sentiment, so easily perhaps bypassed: recall incidents of people not wanting to look at someone in danger, not opening a window to the screams of a woman in Queens being murdered. The many ways we turn away from social responsibility form another cognitive dissonance, which demonstrates the devolution of human contact and social help.

Recall the Milgram experiments (Milgram, 1974), which would not likely make it past any modern-day review board; subjects who were ordered by an authority figure in white coats to give electric shocks to screaming victims (despite their ignorance) that the shocks and screams were faked. The amazing rate of obedience to an instruction to inflict pain was astonishing at that time.

Conformity and compliance, lack of judgment, and loss of individual perspective for the inside of another underlies obedience to authority under these conditions. Despite these caveats, empathy is still a vital form of moral sensibility, and a broad array of moral sentiments (e.g. sympathy/empathy, fairness, duty, self-regulation) have been uncovered that underlie the basis for approach/avoidance social behaviors; sentiments are richly embedded in cephalic information-processing and appraisal systems.

Social attachment is a primary adaptation; evolved sets of neural systems are designed to facilitate social contact. Distinct sets of neurons in diverse regions of the cortex are active when one performs an action and when one watches others do so; this is pristinely shown in studies in macaques (Rolls, 2000). That does not mean that there is not overlap in neurons that fire to mirroring others and in performing the action, it just is so that we come prepared to respond to others. The prosocial systems are tightly linked to moral systems (Adolphs, 1999).

At a basic level, imitation is a vital way in which we access the world through learning about others (Meltzoff, 1977). Starting in neonates, imitation is easily incorporated into how we become connected to others. It is a fundamental adaptation and part of the roots of social bonding, and under the right conditions, human imitation presumably can become connected to prosocial sensibilities such as human empathy. Imitation is just a way to get going, though; it does not stop there, and is richly expressed by neocortical activation including frontal and parietal cortex, in addition to striate and amygdala activation (Frith, 2007).

Simulation of an action, obviously, does not mean the same as doing it, but simulation is a vital part of mind reading. What underlies the research on mirror neurons is the rapidity with which we learn to do things and to imitate one another; recent research has linked social imitation to regions (basal ganglia) that underlie many forms of rule governed behavior (Berridge, 2007), including prosocial sentiments linked to cognitive resources.

While biological determinism is a thing of the past, biological vulnerability and comorbidity is not. There are genes that create a vulnerability to decreases in social attachment and social isolation interacting with other genes for other traits (aggression), which can do harm to the long-term viability and well-being of an individual and decrease individual happiness. Prosocial moral sensitivity allows humans to quickly apprehend the moral implications in a social situation depending on context, agency and consequences of one's choices. These sentiments are intrinsically linked to daily social interactions, and there are several regions in the brain which provide a context for social flexibility. One virtue, moral and otherwise, is in "deciding together" (Moreno, 1995). Recognizing the intentions of others is one critical feature of prosocial behavior. This cognitive capacity begins early in ontogeny and is tied in to visual sensibility (Tomasello, 1999). Eye contact, again, recognizing the intentions of others, is compromised in autistic individuals (Baron-Cohen, 1995/2000). Recognizing the intentions is knotted to a broad array of cephalic tissue that underlies perspective-taking and human social judgment (Adolphs, 1999).

But a propensity for prosocial behaviors is small and competitive with the rest or our predilections. A consideration of the prosocial instincts embedded in cephalic bodily responses towards others exists amidst a frail sensibility of human kindness, as well as diverse cognitive predilections amidst human/animal sentiments, and diverse cognitive rules. Diverse regions of the brain tied to social behaviors underlie prosocial sentiments (Adolphs, 1999).

INFORMATION MOLECULES, CRH

Genetic and epigenetic factors interact with environments that select some features over others in which vulnerable traits interact with environments (Keverne and Curley, 2008) for development of well-being and meaningful social interactions. Genes for CRH or glucocorticoid receptor expression, for instance, set the conditions for social withdrawal in some social contexts (Smoller *et al.*, 2005), though in the case of CRH, the gene takes into account a very small feature of the variance. Put that together with genes for the serotonin form linked to social anxiety and with dopaminergic genes linked to social avoidance (Schmidt *et al.*, 2007), and then a story of social vulnerability begins to build by adding on comorbid factors.

Thus, temperamental differences reflected perhaps in perinatal and early development periods may orient the child too much in one direction (Ahnert *et al.*, 2004). Higher levels of cortisol are not a good thing if they remain high for long periods. The regulation of the internal milieu is compromised, spurring anxious behaviors and self-medication; diverse genetic features contribute to temperament (Kagan, 2002). Again promoter polymorphisms in, for instance, CRH (Smoller *et al.*, 2005), can alter the balance for the expression of social-related fear or social withdrawal.

Genetic polymorphic changes, gene duplication in core neuro-transmitters, such as dopamine and serotonin, can in suitable envir-onments either reveal devolution of function or continued adaptation. However, environment always matters; nothing good happens with-out normal development, and the biology can be fetal and postnatal in which end organ systems will be affected (Barker, 2004). Many factors, including social adversity, nutrition, preeclampsia, and infectious dis-eases create the conditions for low birth weight babies, and for longer-term consequences for human health and well-being (Barker, 2004).

Importantly, peptides such as oxytocin, vasopressin and CRH in the placenta, which is an ancient organ tied importantly to our evolution and to carrying every information molecule in every organ of the human body over the pregnancy period, influences the length and outcome of the pregnancy. Peptides such as oxytocin and CRH (to name just two) have profound effects on the length of gestation and developmental outcome. Oxytocin is an ancient hormone tied to fluid regulation, successful courtship across many species, and success-ful gestation by regulation of estrogen by the induction of oxytocin expression in the placenta, and perhaps the brain. Oxytocin and vaso-tocin or vasopressin, tied to fluid regulation, are bound to postpartum social attachment in mother–infant relationships, social contact, and territorial defense; differences in the vasopressin 1a receptor underly diverse behaviors (Donaldson and Young 2008).

OXYTOCIN/VASOPRESSIN

Oxytocin is a richly endowed hormone with a phylogenetically inter-esting history. It is bound to many forms of biological regulation

Table C.1. *Possible role of oxytocin in causing long-term benefits of positive social interaction (Uvnäs-Moberg, 1998)*

OXYTOCIN
↓
Positive social interaction
↓
Activation of sensory afferents by nonnoxious stimuli
↓
OXYTOCIN
↓
Stimulation of antistress effects and growth
↓
(Stimulation of attachment or bonding)
↓
Repetitive sensory, mental andconditioned stimuli of nonnoxious type
↓
OXYTOCIN
↓↓↓
Sustained antistress effects and stimulation of growth
↓↓↓
Promotion of health

important over the life cycle (Carter *et al.*, 1997/1999; Neuman, 2008). Social attachment has short- and long-term regulatory functions essential for the whole of one's lifetime. One adaptation is social contact; premature babies huddled together have a decreased recovery period versus those in isolated hospital units. Information molecules such as oxytocin promote social contact; but a context congenial for this has to be somewhat present, or constructed. Neuropeptides are more specific than classical neurotransmitters with regard to behavior. But a context must be permissive for the behavioral options. Oxytocin will not create love – oxytocin is not the hormone of love (though it will probably be marketed that way some day) but it is tied to human social attachment (Table C.1).

Oxytocin and vasopressin are both ancient genes, with a rich homological history, some of which are depicted in Figure C.3 (Donaldson and Young, 2008).

However, a slight change in the receptor structure (in this case the V1 receptors) can perhaps have a profound impact on our social behaviors, both affiliative, during normal behaviors, and devolutionary, in autism; genetic polymorphism figures importantly for social behaviors (Donaldson and Young, 2008).

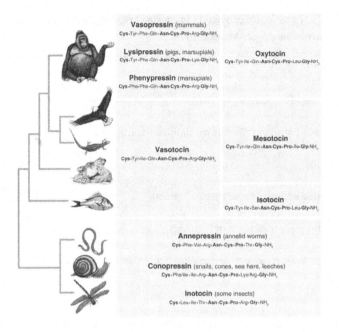

Figure C.3   Oxytocin and vasopressin homologs (from Donaldson, Z. R., & Young, L. J. (2008). Oxytocin, vasopressin and the neurogenetics of sociality. *Science*, **322**, 900–4. Reprinted with permission from AAAS).

PROLACTIN

For those species in which parental behaviors occur in both genders, neuropeptides, such as oxytocin and prolactin, are importantly involved. And, of course, the fact that central prolactin is linked to child rearing is illustrated by the presence of higher levels of prolactin in males of diverse species, such as in the new world monkey and the common marmoset (Mota and Sousa, 2000), who contribute to child rearing. Interestingly, avian parenting, in which both males and females contribute much more toward the child rearing, may have emerged with dinosaurs, suggesting an even more ancient role for hormones such as prolactin.

Information molecules such as prolactin contribute towards our adaptation to social surroundings; our evolution is towards social complexity in which social cooperation and cephalic expansion are fundamental factors (Barton, 2004). But hormones of activation and cooperation or deception are bound to regulatory information molecules. Cortisol, a hormone that wakes us up and lies at the heart of the organization of action, and prolactin, a hormone of social attachment, are also linked to the transduction of light to dark in many species (Wehr *et al.*, 1993) and the quiescence that underlies our social behavioral capabilities. A circadian clock is bound to hormones such as cortisol and prolactin.

Perhaps it is not surprising that the hormone prolactin has been putatively linked to transition periods between light and dark phases of the circadian regulatory systems; phases of dusk (for instance, the quiet time before sleep) and prolactin may be bound to the transition period, along with the social attachment and social conformity that emerges under these conditions. Interestingly, meditators have a tendency to secrete prolactin (Wehr and Schulkin, unpublished observations), switching the cortical expression towards positive affect in humans (Davidson *et al.*, 2003); the sitting behavior is reminiscent of the brooding behaviors of birds sitting over their eggs.

SOCIAL ALLOSTASIS AND WELL-BEING

Reaching out to others is but one prosocial response that we all share, even though it varies across cultures in its expression. The life blood of humanity, after all, is our social bonds and the way we enjoy and manage our solitariness (Sterling, 2004).

Allostatic overload, for instance, by exaggerated levels of cortisol, sets the stage for a wellspring of devolution of function and exaggerated breakdown of bodily tissue (McEwen, 1998); one ameliorative factor is social meaning; contact. Our brains are designed for social cooperative behaviors and social deception, amongst other forms of contact. Social contact, meaningful close relationship, is an important factor in well-being; sleep, for instance, is linked to meaningful social contact (McEwen, 2006).

In fact, one recurrent feature is that lending a hand to others, social responsivity to others, acts to placate the response to threat. Moreover, chronic arousal, exposure to endless seductive meaningless ploys, is a huge drain on cephalic and systemic resources; meaningful social contact, conversely, is an ameliorative biological adaptation (Jaspers, 1913/1997).

The important role of satisfaction in the organization of action and its devolution has long been noted. We are bombarded by diverse signals that are tied into satisfaction in a culture of broad-based consumption; we are the consuming society, with an ever expanding capacity for doing so. After all, the neural/physiological machinery is well endowed, and all the pull is for consumption; witness the explosion of obesity knotted to the availability of junk foods and a reduction of natural exercise.

Well-being is linked to many factors including exercise, diet, health, temperament and life events, all of which contribute to allostatic regulation and allostatic breakdown. Allostatic overload is a chronic social fact; the origins of the concept of allostasis, after all, emphasized the social milieu, and social status is important to vulnerabilities. Social interventions, in which people in an aging population can help others through meaningful contact and impart what Dewey called "funded knowledge" are both socially important and important to the individual giving the knowledge.

Of course, health is a normative goal set in a social context; being positive cannot cure cancer, but it can sustain one in hard times. There is no cure or panacea, but there are correlations of meaning; social contact can indeed lessen allostatic overload. Social duress under social inequities facilitates cortisol increases, but friendly social contact decreases excesses in cortisol levels (Gunnar et al., 1989). In an aging population such as ours, having others reach out as an "experience core" (Fried et al., 2004) serves everyone involved. And many factors go into well-being, including genes, parental/social developmental and long-term relationships, key experiences and health.

However, an endless seduction consumes culture with little actual reward and only fleeting gratifications lead to a devolution of human well-being. The stick metaphor is a tonic indicator for a warning about dysregulation and the longer-term view; a carrot metaphor works for shorter-term satisfaction (Peter Sterling, pers. comm., 2005).

A devolution of satisfaction and constant disharmony results from a social world in which, using another metaphor, the "lights are always on," evoking chronic arousal. What can happen from chronic arousal is diminished satisfaction in a chronically enhanced incentive social milieu: too little time to appreciate, too much to look at, too many seductions. In other words, devolution of satisfaction is coupled with the loss of taking one's time. No doubt, this has the ring of truth; we can all see a decline of social contact and meaning in our world.

Less trust amongst us, more exposure to strangers, fewer family ties, greater unease – of course not all of us experience the world this way, and the social milieu and the larger ecological context has always been littered with unease, uncertainty and endless danger. Furthermore, our cognitive capabilities are rarely consistent and often short-sighted (Kahneman *et al.*, 1999). Amidst these events are mean-ingful social bonds, and thus, well-being (Jaspers, 1913/1997). Evolution put a premium on social attachment, gaining a foothold into the world from being guided and connected to others; the cooperative tendencies coexist amidst the competitive predilections, the desire to live.

Social contact is at the heart of ontogenetic development, a long noted piece of epistemological history, differently expressed across diverse cultures. Family and group structure through meaningful con-tact are an essential for our mental health. Supportive social contact is not an absolute prophylactic but a helpful ameliorative in combating disease and breakdown; intermittent unpredictable aversive events are a long known production of pathology (e.g. gastric), increasing allo-static load. Loneliness is a devolution of the ameliorative social factor; meaningful labor is at the heart of the human condition and under-lies recovery and maintenance of well-being. In other conditions that might be favorable to neuroendocrine systems, social isolation exacer-bates vulnerabilities across diverse species (Cacioppo *et al.*, 2006).

Unremitting social distress is cumulative, and facilitates the aging process and increases the allostatic load. Allostatic load is one predictive factor in aging; age, health and economic disparities are all functionally related to allostatic overload. Small everyday transac-tions, ranging from forgiveness to less interpersonal conflict, good exercise and diet, and perhaps a positive sensibility can have, in con-trast, ameliorative results. Small things matter, little gifts that ameli-orate negative conditions; helping others is indeed its own reward.

Many of the prosocial behaviors and the evaluative processes in cephalic systems are anticipatory and not only reactive to events. Information molecules, such as oxytocin, CRH and vasopressin, under-lie diverse forms of anticipatory behaviors. The concept of "allostasis" is in part to take account of anticipatory control (Sterling and Eyer, 1988) amidst diverse forms of cognitive adaptation underlying this regulatory adaptation that supports social contact and internal milieu (Schulkin, 2003). We are social animals as Aristotle noted; anticipa-tory mechanisms evolved with the social forms of adaptations, taking account of one another, foraging for food, building alliances of social cooperation, deception and confrontation in group formation.

Cephalic or brain organization is vital in the organization of social formation. One putative mechanism tied to anticipatory regula-tion is that of allostasis, a term for the physiology of change. In fact, the view of the brain as a simple chain of reflexes has given way to a view of the brain with diverse central programs rich in syntactical scope and emboldened by a diversity of cognitive capacties across the central nervous expression, particularly in the case of our species. It is not surprising that a conception of probability of generating one response or another has became a better way the brain selects a response amongst competing possible behaviors. One set of mechanisms for this are the diverse forms of rhythmic cephalic generators that help sets the condition for the selection.

Thus the brain is an active organ generating central rhythmic forms of behavior bound to real world events in which objects, or what J. J. Gibson (1979) called "affordance" objects with value to us, have high selective value for which the nervous system has been recruited. Diverse chemical signals in the brain are under circadian rythmicity, for example, which set the conditions for search or foraging behaviors, attachment behaviors, etc. Allostatic regulation is the lifeblood of cephalic innervations of adaptive physiological and behavioral responses. The design of the brain is in the discernment of affordable objects that serve the multiplicity of functions amidst brain mechanisms that underlie approach/avoidance gradients; and the sentiments that link us and that sustain human contact with diverse variation in culture. We all know that wealth is far from happiness, but it does not hurt. The long litany of prosocial emotions contribute, essentially, to well-being.

Of course, we know that a sense of well-being can be found in generosity, in reaching out to others, and that it is linked to moral ethos. Aristotle distinguishes the short-sighted well-being of the senses from the well-being tied to a meaningful life. In modern form, the difference between the hedonic treadmill, the computational mistakes and expectations of the decision sciences of the senses (Kahneman et al., 1999) and the longer term of what constitutes a meaningful, eudemonic life is a distinction with which we are, or should all be familiar (Dewey, 1925/1989). Diverse cephalic systems are implicated in this process. In the end trust, meaningful labor, and social contact amidst human kindness and social solidarity are essential for our well-being.

References

Additional references and citations may be found at the website for this book, www.cambridge.org/9780521509923.

Abbott, D. H., Keverne, E. B., Bercovitch, F. B., et al., (2002). Are subordinates always stressed? *Hormones and Behavior*, **54**, 78–93.

Adolphs, R. (1999). Social cognition and the human brain. *Trends in Cognitive Sciences*, **3**, 469–479.

Adolphs, R., Denburg, N. L., and Tranel, D. (2001). The amygdala's role in long-term declarative memory for gist and detail. *Behavioral Neuroscience*, **115**, 983–992.

Adolphs, R., Gosselin, F., Buchanan, T. W., Tranel, D., Schyns, P., and Damasio, A.R. (2005). A mechanism for impaired fear recognition after amygdala damage. *Nature*, **433**, 68–72.

Aggleton, J. (1992/2000). *The amygdala*. Oxford: Oxford University Press.

Ahnert, L., Gunnar, M. R., Lamb, M. E., and Barthel, M. (2004). Transition to childcare: Associations with infant-mother attachment, infant negative emotion, and cortisol elevations. *Child Development*, **75**, 639–650.

Aiello, L. C, and Dunbar, R. I. M (1993). Neocortex size, group size, and the evolution of language. *Current Anthropology*, **34** (2), 184–193.

Almond, R. E. A., Brown, G. R., and Keverne, E. B. (2006). Suppression of prolactin does not reduced infant care by parentally experimed male common marmosets. *Hormones and Behavior*, **49**, 673–680.

Aristotle (1962). *The nicomachean ethics*. New York: Macmillan.

Arnold, A. P. (2002). Concepts of genetic and hormonal induction of vertebrate sexual differentiation in the twentieth centrury with special reference to the brain. In D. W. Pfaff, et al., (Eds.), *Hormones, brain and behavior*. New York: Elsevier Press.

Ball, G. F., and Balthazart, J. (2002). Neuroenocrine mechanisms regulating reproductive cycles and reproductive behavior in birds. In D. W. Pfaff (Ed.), *Hormones, brain and behavior*. New York: Academic Press.

Barger, N., Stefanacci, L., and Semendeferi, K. (2007). A comparative volumetric analysis of the amygdaloid complex and basolateral division in the human and ape brain. *American Journal of Physical Anthropology*, **134**, 392–403.

Barker, D. J. (2004). The developmental origins of well-being. *Phil. Trans. R. Soc. London. B*, **359**, 1359–1366.

Baron-Cohen, S. (1995/2000). *Mindblindness*. Cambridge, MA: MIT Press.

Baron-Cohen, S., Lutchmaya, S., and Knickmeyer, R. (2004). *Prenatal tesosterone in mind*. Cambridge: MIT Press.

Barton, R. A. (2004). Binocularity and brain evolution in primates. *PNAS*, **101**, 10113–10115.

Bauman, D. E. (2000). Regulation of nutrient portioning during lactation: homeostasis and homeorhesis revisisted. In P. J. Cronje (Ed.), *Ruminant physiology*. New York: CAB Publishing.

Beaton, E. A., Schmidt, L. A., Schulkin, J., Antony, M. M., Swinson, R. P., and Hall, G. B. (2008). Different neural responses to stranger and personally familiar faces in shy and bold adults. *Behav Neurosci*, **122**, 704–9.

Bechara, A. (2005). Decision-making, impulse control and loss of willpower to resist drugs: a neurocognitive perspective. *Nature Neuroscience*, **8**, 1458–1463.

Bentley, P. J. (1982). *Comparative vertebrate endocrinology*. Cambridge: Cambridge University Press.

Berntson, G. G., and Cacioppo, J. T. (2000). From homeostasis to allodynamic regulation. In J. T. Cacioppo, L. G. Tassinary, and G. G. Berntson (Eds), *Handbook of psychophysiology*. Cambridge: Cambridge University Press.

Berridge, K. C. (2007). The debate over dopamine's role in reward: The case for incentive salience. *Physiology and Behavior*, **191**, 391–431.

Berridge, K. C. (2004). Motivation concepts in behavioral neuroscience. *Physiology and Behavior*, **81**, 179–209.

Berridge. K. C., Grill, H.J., and Norgren, R. (1981). The relation of consummatory resopnses and preabsorptive insulin release to palatability and learned taste aversions. *Journal of Comparative and Physiological Psychology*, **95**, 363–382.

Bindra, D. (1974). A motivational view of learning, performance, and behavior modification. *Psychological Review*, **81**, 199–213.

Bodnar, R. J., Commons, K., and Pfaff, D. W. (2002). *Central neural states relating sex and pain*. Baltimore, MD: Johns Hopkins University Press.

Botvinick, M. M., Braver, T. S., Barch, D. M., Carter, C. S., and Cohen, J. D. (2001). Conflict monitoring and cognitive control. *Psychol Rev*, **108**, 624–52.

Bridges, R. S., and Mann, P. E. (1994). Prolactin-brain interactions in the induction of maternal behavior in rats. *Psychoneuroendocrinology*, **19**(5–7), 611–622.

Broca, P. (1878). Anatomic comparee des ciconvolutions le grad lbe limbique et la scissure limbique dans la serie des mammiferes. *Review of Anthropology*, **1**, 385–498.

Brown, P., and Marsden, C. D. (1998). What do the basal ganglia do? *The Lancet*, **351**, 1801–1804.

Buntin, J. D., Ruzycki, E, and Witebsky, J. (1992). Prolactin receptors in dove brain: Autoradiographic analysis of binding characteristics in discrete brain regions and accessibility to blood-borne prolactin. *Neuroendocrinology*, **57**, 738–750.

Burkhardt, R.W. (1975/1997). *Lamarck and evolutionary biology*. Cambridge, MA: Harvard University Press.

Byrne, R. W. and Bates, L. A. (2007). Sociality, evolution and cognition. *Current Biology*, **17**, R714–R723.

Byrne, R. W., and Corp, N. (2004). Neocortex size predicts deception rate in primates. *Proc. R. Soc.* **271**, 1693–1699.

Cacioppo, J. T., Hawkley, L., and Ernst, J. M., et al. (2006). Loneliness within a nomological net: an evolutionary perspective. *Joural of Research in Personality*, **40**, 1054–1085.

Cacioppo, J. T., Visser, P. S., and Pickett, C. L. (2006). *Social neuroscience*. Cambridge: MIT Press.

Cajal, S. R. (1906). The structure and connexions of neurons. In *Nobel Lectures, Physiology or Medicine 1901–1921*. New York: Elsevier, 220–253.

Cannon, W. B. (1916/1929). *Bodily changes in pain, hunger, fear and rage*. New York: Appleton and Co.

Carey, S. (2009). *The origin of concepts*. Oxford: Oxford University Press.

Carter, C. S. (2007). Sex differences in oxytocin and vasopressin: implications for autism spectrum disorders? *Behavioural Brain Research*, **176**, 170–186.

Carter, C. S., Lederhendler, I. L., and Kirkpatrick, B. (1997/1999). *The integrative neurobiology of affiliation*. Cambridge, MA: MIT Press.

Cavigelli, S. A., Dubovick, T., Levash, W., Jolly, A., and Pitts, A. (2003). Female dominance status and fecal corticoids in a cooperative breeder with low reproductive skew: ring-tailed lemurs. *Hormones and Behavior*, **43**, 166–179.

Charuvastra, A., and Cloitre, M. (2008). Social bonds and posttraumatic stress disorder. *Annu Rev Psychol*, **59**, 301–328.

Chauvet, J., Hurpet, D., Michel, G., Chauvet, M. T., Carrick, F. N., and Acher, R. (1985). The neurohypophysial hormones of the egg-laying mammals: identification of arginine vasopressin in the platypus (Ornithorhynchus anatinus). *Biochem Biophys Res Commun*, **127**, 277–282.

Cheney, D. L., and Seyfarth, R. M. (1990). *How monkeys see the world*. Chicago, IL: University of Chicago Press.

Cheney, D. L., and Seyfarth, R. M. (2007). *Baboon metaphysics*. Chicago, IL: University of Chicago Press.

Choleris, E., Ogawa, S., Kavaliers, M., Gustafsson, J. A., Korach, K. S., Muglia, L. J., and Pfaff, D. W. (2006). Involvement of estrogen receptor alpha, beta and oxytocin in social discrimination: A detailed behavioral analysis with knockout female mice. *Genes Brain Behav*, **5**, 528–539.

Connor, R. C. (2007). Dolphin social intelligence: Complex alliance relationships in bottlenose dolphins and a consideration of selective environments for extreme brain size evolution in mammals. *Philos Trans R Soc Lond B Biol Sci*, **362**, 587–602.

Cook C.J. (2002). Glucocorticoid feedback increases the sensitivity of the limbic system to stress. *Physiol. Behav.* **75**: 455–464.

Craig, W. (1918). Appetites and aversions as constituents of instincts. *The Biological Bulletin* **34**, 91–107.

Crews, D. (2008). Epigenetics and its implications for behavioral endocrinology. *Front in Neuroend*, **29**, 344–357.

Crockett, M. J., Clark, L., Tabibnia G., Lieberman, M. D., and Robbins, T. W. (2008). Serotonin modulates behavioral reactions to unfairness. *Science*, **320**, 1739.

Curley, J. P., and Keverne, E. B. (2005). Genes, brains and mammalian social bonds. *Trends Ecol Evol*, **20**, 561–567.

Dallman, M. F., Akana, S. F., Cascio, C. S., Darlington, D. N., Jacobson, L., and Levin, N. (1987). Regulation of ACTH secretion: Variations on a theme of B. *Rec Prog Horm Res*, **43**, 113–173.

Dallman, M. F., Pecoraro, N., Akana, S. F., *et al.* (2003). Chronic stress and obesity: a new view of "comfort food". *Proc Natl Acad Sci*, **100**, 11696–11701.

Damasio, A. R. (1996). The somatic marker hypothesis and the possible functions of the prefrontal cortex. *Philosophical Transactions of the Royal Society of London*, **354**, 1413–1420.

Damasio, A. R., Tranel, D., and Damasio, A. R. (1990). Individuals with sociopathic behavior caused by frontal damage fail to respond autonomically to social stimuli. *Behavioral Brain Research*, **41**, 81–94.

Darwin, C. (1859/1958). *The origin of species*. New York: Mentor Book.

Darwin, C. (1874). *The Descent of Man and Selection in Relation to Sex*. 2nd edn. Reprinted 1896, New York: D. Appleton & Co.

Davidson, R. J., Kabat-Zinn, J., Schumacher, J., et al., (2003). Alterations in brain and immune function produced by mindfulness meditation. *Psychosomatic Medicine*, **65**, 564–570.

Davis, M., Walker, D. L., and Lee, Y. (1997). Amygdala and bed nucleus of the stria terminalis: Differential roles in fear and anxiety measured with the acoustic startle reflex. *Philos Trans R Soc Lond B Biol Sci*, **352**, 1675–1687.

Davis, M., Walker, D. L., Miles, L., and Grillon, C. (2010). Phasic vs. sustained fear in rats and humans: Role of the extended amygdala in fear vs. anxiety. *Neuropsychopharmacology*, **35**, 105–135.

De Kloet, E.R. (1991). Brain corticosteroid receptor balance and homeostatic control. *Front Neuroendocrinol*, **12**, 95–164.

De Kloet, E. R., Jeols, M., and Holsboer, F. (2005). Stress and the brain: From adaptation to desease. *Nature*, **6**, 463–472.

De Vries, D.J. and Miller, M.A. (1998). Anatomy and function of extrahypothalamic vasopressin systems in the brain. *Progress in Brain Research*, **119**, 3–20.

De Waal, F., and Lanting, F. (1997). *Bonobo: The forgotten ape*. Berkeley, CA: University of California Press.

Delgado, M. R., Nearing, K. I., LeDoux, J. E., and Phelps, E. A. (2008). Neural circuitry underlying the regulation of conditioned fear and its relation to extinction. *Neuron*, **11**, 829–838.

Denton, D. (2005). *The primordial emotions*. Oxford: Oxford University Press.

Denton, D. (1982). *The hunger for salt*. Berlin: Springer-Verlag.

Denver, R. J. (1998). Hormonal correlates of environmentally induced metamorphosis in the western spadefoot toad, Scaphiopus hammondii. *General and Comparative Endocrinology*, **110**, 326–336.

Denver, R. J. (2009). Endocrinology of complex life cycles: amphibians. In D. Pfaff, A. Arnold, A. Etgen, S. Fahrbach, R. Moss and R. Rubin (Eds), *Hormones, brain and behavior*, 2nd edn. Amsterdam: Elsevier.

Dethier, V. G., and Stellar, E. (1961/1970). *Animal behavior*. Englewood Cliffs, NJ: Prentice Hall.

Dewey, J. (1925/1989). *Experience and nature*. LaSale, IL: Open Court Press.

Diamond, A. (2001). A model system for studying the role of dopamine in the prefrontal cortex during early development of humans: early and continuously treated phenylketonuria. In C. A. Nelson and M. Luciana (Eds.), *Handbook of developmental cognitive neuroscience*. Cambridge, MA: MIT Press, 433–472.

Diamond, A., and Amaso, D. (2008). Contributions of neuroscience to our understanding of cognitive development. *Curr Dir Psychol Sci*, **17**, 136–141.

Dingle, H. (2002). Hormonal mediation of insect life histories. In D. W. Pfaff, et al., (Eds.), *Hormones, brain and behavior*. New York: Elsevier.

Dixson, A. F., and George, L. (1982). Prolactin and parental behaviour in a male New World primate. *Nature*, **299**, 551–553.

Dobzhansky, T. C. (1962). *Mankind evolving*. New Haven, CT: Yale University Press.

Dolan, R. (2007). The human amygdala and orbital prefrontal cortex in behavioral regulation. *Phil Trans R Soc*, **362**, 787–789.

Donald, M. (1991). *Origins of the modern mind*. Cambridge, MA: Harvard University Press.

Donaldson, Z. R., and Young, L. J. (2008). Oxytocin, vasopressin and the neurogenetics of sociality. *Science*, **322**, 900–904.

Donley, M. P., Schulkin, J., and Rosen, J. B. (2005). Glucocorticoid receptor antagonism in the basolateral amygdala and ventral hippocampus interferes with long-term memory of contextual fear. *Behavioural Brain Research*, **164**, 197–205.

Drago, F., D ' Agata, V., Iacona, T., *et al.* (1989). Prolactin as a protective factor in stress-induced biological changes. *Journal of Clinical Laboratory Analysis*, **3**, 340–344.

Dunbar, R. I. M. (1992). Neocortex size as a constraint on group size in primates. *J Hum Evol*, **22**, 469–493.

Dunbar, R. I. M. (1996). *Grooming, gossip, and the evolution of language*. Cambridge, MA: Harvard University Press.

Dunbar, R. I. M. (2003). The social brain. *Annual Review of Anthropology*, **32**, 163–181.

Dunbar, R. I. M., and Shultz, S. (2007). Evolution in the social brain. *Science*, **317**, 1344–1347.

Eldridge, N. (1985). *Unfinished synthesis*. Oxford: Oxford University Press.

Emery, N. J. and Amaral, D. G. (2000). The role of the amygdale in primate social cognition. In R. D. Lane and L. Nadel (Eds), *Cognitive neuroscience of emotion*. New York: Oxford University Press.

Engh, A. L., Beehner, J. C., Bergman, T. J., *et al.*, (2006). Female hierarchy instability, male immigration and infanticide increase glucocorticoid levels in female chacma baboons. *Animal Behavior*, **71**, 1227–1337.

Epstein, A. N. (1991). Neurohormonal control of salt intake in the rat. *Brain Research Bulletin*, **27**, 315–320.

Erickson, K., Drevets, W. C., and Schulkin, J. (2003). Glucocorticoid regulation of diverse cognitive functions in normal and pathological emotional states. *Neuroscience and Biobehavioral Reviews*, **27**, 233–246.

Erickson, K., Gabry, K. E., Lindell, S., *et al.* (2005). Social withdrawal behaviors in nonhuman primates and changes in neuroendocrine and monoamine concentrations during a separation paradigm. *Dev Psychobiol*, **46**, 331–339.

Erickson, K., Thorsen, P., Chrousos, G., Grigoriadis, D. E., Khongsaly, O., and Schulkin, J. (2001). Preterm birth: associated neuroendocrine, medical and behavioral risk factors. *Clinical Endocrinology and Metabolism*, **86**, 2544–2552.

Evans, P. D., Mekel-Bobrov, N., Vallender, E. J., Hudson, R. R., and Lahn, B. T. (2006). Evidence that the adaptive allele of the brain size gene microcephalin introgressed into Homo sapiens from an archaic Homo lineage. *Proc Natl Acad Sci USA*, **103**, 18178–18183.

Febo, M., Numan, M., and Ferris, C. F. (2005). Functional magnetic resonance imaging shows oxytocin activates barin regions associated with mother-pup bonding during suckling. *Journal of Neuroscience*, **25**, 11637–11644.

Ferguson, J. N., Young, L. J., Hearn, E. F., Matzuk, M. M., Insel, T. R., and Winslow, J. T. (2000) Social amnesia in mice lacking the oxytocin gene. *Nature Genetics*, **25**, 284–288

Ferguson, J. N., Aldag, J. M., Insel, T. R., and Young, L. J. (2001). Oxytocin in the medial amygdala is essential for social recognition in the mouse. *J of Neuroscience*, **21**, 8278–8265.

Festinger, L. (1957). A theory of cognitive dissonance. Palo Alto: Stanford University Press.

Fitzsimons, J. T. (1979). *The physiology of thirst and sodium appetite*. Cambridge: Cambridge University Press.

Fliessbach, K., Weber, B., Trautner, P., Dohmen, T., Sunde, U., Elger, C. E., and Falk, A. (2007). Social comparison affects reward-related brain activity in the human ventral striatum. *Science*, **318**, 1305–1308.

Foley, R. (2006). The emergence of culture in the context of hominin evolutionary patterns. In S. C. Levinson and P. Jaisson (Eds.), *Evolution and culture*. Cambridge, MA: MIT Press.

Foley, R., and Lahr, M. M. (2004). Human evolution writ small. *Nature*, **431**, 1043–1044,

Francis, D. D., Diorio, J., Liu, D., and Meaney, M. J. (1999). Nongenomic transmission across generations of meternal behavior and stress responses in the rat. *Science*, **286**, 1155–1158.

Francis, D. D., Young L. J., Meaney, M. J., and Insel, T. R. (2002). Naturally occurring differences in maternal care are associated with the expression of oxytocin and vasopression (V1a) receptors: Gender differences. *J of Neuroendocrinology* **14**, 349–353.

Frederick, S., and Loewenstein, G. (1999). Hedonic adaptation. In D. Kahneman, E. Diener, and N. Scharz (Eds.), *The foundations of hedonic psychology*. New York: Russell Sage Foundation.

Fried, L. P., Carlson, M. C., Freedman, M., Frick, K. D., Glass, T. A., Hill, J., et al. (2004). A social model for health promotion for aging population. *J. of Urban Health: Bulletin of the New York Academy of Medicine*, **81**, 64–77.

Frim, D. M., Emanuel, R. L., Robinson, B. G., Smas, C. M., Adler, G. K., and Majzoub, J. A. (1988). Characterization and gestational regulation of corticotropin-releasing hormone messenger RNA in human placenta. *J Clin Invest*, **82**, 287–292.

Frith, C. D. (2007) The social brain? *Philos Trans R Soc Lond B Biol Sci*, **362**, 671–678.

Galef, B. G., and Whiskin, E. E. (2000). Social influences on the amount of food eaten by Norway rats. *Appetite*, **34**, 327–332.

Gallagher, M., and Holland, F. C. (1994). The amygdala complex: Multiple roles in associative learning and emotion. *Proc. Of the Nat. Acad. Of Sci.*, **91**, 11771–11776.

Gallistel, C. R. (1980). *The organization of action: A new synthesis*. Hillsdale, NJ: Lawrence Erlbaum.

Garcia, V., Jouventin, P., and Mauget, R. (1996). Parental care and the prolactin secretion pattern in the king penguin. *Hormones and Behavior*, **30**, 259–265.

Gibson, J. J. (1979). *The ecological approach to visual perception*. Boston, MA: Houghton Mifflin.

Gigerenzer, G. (2000). *Adaptive thinking, rationality in the real world*. New York: Oxford University Press.

Gluckman, P., and Hanson, M. (2005). *The fetal matrix*. Cambridge: Cambridge University Press.

Goland, R. S., Wardlaw, S. L., Stark, R. I., Brown, L. S., and Frantz, A. G. (1986). High levels of corticotropin-releasing hormone immunoreactivity in maternal and fetal plasma during pregnancy. *J Clin Endocrinol Metab*, **63**, 1199–1204

Goldman-Rakic, P. S., Leranth, C., Williams, S. M., Mons, N., and Gerrard, M. (1989) Dopamine synaptic complex with pyradidal neurons in primate cerebral cortex. *Proc. Nat. Acad. Sci.*, **86**, 9015–9019.

Gould, E., McEwen, B. S., Tanapat, P., Galea, L. A., and Fuchs, E. (1996). Neurogenesis in the dentate gyrus of the adult tree shrew is regulated by psychosocial stress and NMDA receptor activation. *J Neurosci*, **17**, 2492–2498.

Gould, S. J. (2002). *The structure of evolutionary theory*. Cambridge, MA: Harvard University Press.

Goymann, W, and Wingfield, J. C. (2004). Allostatic load, social status and stress hormones: the costs of social status matter. *Animal Behaviour*, **67**, 591–602.

Greene, J. D., Morelli, S. A., Kowebgerg, K., *et al.*, (2008). Cognitive load selectively interferes with utilitarian moral judgment. *Cognition*, **107**, 1144–1154.

Greene, J. D., Nystrom, L. E., Engell, A. D., Darley, J. M., and Cohen, J. D. (2004). The neural bases of cognitive conflict and control in moral judgment. *Neuron*, **44**, 389–400.

Greene, J.D., Sommerville, R. B., Nystrom, L. E., Darley, J. M., and Cohen, J. D. (2001). An fMRI investigation of emotional engagement in moral judgment. *Science*, **293**, 2105–2108,

Griffin, D. R. (1958). *Listening in the dark*. New Haven, CT: Yale University Press.

Gubernick, D. J., Winslow, J. T., Jensen, P., Jeanotte, L., and Bowen, J. (1995). Oxytocin changes in males over the reproductive cycle in the monogamous, biparental California mouse, Peromyscus californicus. *Horm Behav*, **29**, 59–73.

Gunnar, M. R., and Davis, E. P. (2001). The developmental psychobiology of stress and emotion in early childhood. In R. M. Lerner, M. A. Easterbrooks, and J. Mistry (Eds.), *Comprehensive handbook of psychology, vol 6, Developmental psychology*. New York: Wiley.

Gunnar, M. R., Mangelsdorf, S., Larson, M., and Hertsgaard, L. (1989). Attachment, temperament, and adrenocortical activity in infancy: a study of psychoendocrine regulation. *Development Psychology*, **25**, 355–363.

Gunterkun, O. (2005). The avian 'prefrontal cortex' and cognition. *Curr Opin Neurobiol*, **15**, 686–693.

Hammock, E. A., and Young, L. J. (2004). Functional microsatellite polymorphism associated with divergent social structure in vole species. *Mol Biol Evol*, **21**, 1057–1063.

Hariri, A. R., Mattay, V. S., Tessitore, A., *et al.* (2002). Serotonin transporter genetic variation and the response of the human amygdala. *Science*, **297**, 400–403.

Heckel, G., and Fink, S. (2008). Evolution of the arginine vasopressin 1a receptor and implications for mammalian social behavior. *Progress in Brain Research*, **170**, 321–330.

Heinrichs, M., Baumgartner, T., Kirschbaum, C., and Ehlert, U. (2003). Social support and oxytocin interact to suppress cortisol and subjective responses to psychosocial stress. *Biol Psychiatry*, **54**, 1389–1398.

Herbert, J. (1993). Peptides in the limbic system: neurochemical codes for co-ordinated adaptive responses to behavioral and physiological demand. *Prog Neurobiol*, **41**, 723–791.

Herbert, J., and Schulkin J. (2002). Neurochemical coding of adaptive responses in the limbic system. In D. Pfaff (Ed.), *Hormones, brain and behavior*. New York: Elsevier.

Herman, J. P., Dolgas, C. M., and Carlson, S. L. (1998). Ventral subiculum regulates hypothalamic-pituitary-adrenocortical and behavioral responses to cognitive stressors. *Neuroscience*, **86**, 449–459.

Herman, J. P., Figueiredo, H., Nueller, N. K., *et al.* (2003). Central mechanisms of stress integration. *Frontiers in Neuroendocrinology*, **24**, 151–180.

Herrick, C. J. (1926/1963). *Brain in rats and men*. New York: Hafner Publishers.

Herrmann, E., Call, J., Hernadez-Lioreda, M. V. *et al.*, (2007) Humans have evolved specialized skills of social cognition. *Science* **317**, 1360–1366.

Hofer, M. A. (1973). The role of nutrition in the physiological and behavioral effects of early maternal separation on infant rats. *Psychosom Med*, **35**, 350–359.

Hollander, E., Novotny, S., Hanratty, M., Yaffe, R., DeCaria', C. M., Aronowitz, B. R., and Mosovich, S. (2003). Oxytocin infusion reduces repetitive behaviors in adults with autistic and Asperger's disorders. *Neuropsychopharmacology*, **28**, 193–198.

Holliday, R., and Ho, T. (1998). Evidence for gene silencing by endogenous DNA methylation. *Proc Natl Acad Sci USA*, **95**, 8727–8732.

Hume, D. (1984). *A treatise of human nature*. New York: Penguin Classics. (Original work published 1739.)

Insel, T. R. (1992). Oxytocin-a neuropeptide for affiliation: Evidence from behavioral, receptor autoradiographic, and comparative studies. *Psychoneuroendocrinology*, **17**, 3–35.

Isomura, Y., Ito, U., Akazawa, T., Nambu, A., and Takada, M. (2003). Neural coding of 'attention for action' and 'response selection' in primate anterior cingulated cortex. *Journal of Neuroscience*, **23**, 8002–8012.

Jackson, J. H. (1884/1958). Evolution and disollution of the nervous system. In: *Selected Writings of John Hughlings Jackson*. London: Staples Press.

Jacobowitz, D. M. (1988). Multifactorial control of pituitary hormone secretion: The "wheels" of the brain. *Synapse*, **2**, 86–92.

James, W. (1890/1952). *The principles of psychology*. New York: Dover Press.

Jasnow, A. M., Schulkin, J., and Pfaff, D. W. (2006). Estrogen facilitates fear conditioning and increases CRH mRNA expression in the central nucleus of the amygdala. *Hormones and Behavior* **49**, 197–205.

Jaspers, K. (1913/1997). *General psychopathology, Vol. I and II*. (J. Hoenig and M. W. Hamilton, Trans.). Baltimore, MD: The Johns Hopkins University Press.

Jolly, A. (1999). *Lucy's legacy*. Cambridge: Harvard University Press.

Kagan, J. (1984). *The nature of the child*. New York: Basic Books.

Kagan, J. (2002). *Surprise, uncertainty and mental structure*. Cambridge, MA: Harvard University Press.

Kagan, J., Resnick, J. S., and Snidman, N. (1988). Biological bases of childhood shyness. *Science*, **240**, 167–171.

Kahneman, D., Diener, E., and Schwarz, N. (1999). *Well being: The foundations of hedonic psychology*. New York: Russell Sage Foundation.

Kalin, N. H., Larson, C., Shelton, S. E., and Davidson, R. J. (1998). Asymmetric frontal brain activity, cortisol, and behavior associated with fearful temperament in rhesus monkeys. *Behav Neurosci*, **112**, 286–292.

Kant, I. (1792/1951). *Critique of judgment*. New York: Hafner Press.

Keverne, E. B., and Curley, J. P. (2008). Epigenetics, brain evolution and behaviour. *Front Neuroendocrinol*, **29**, 398–412.

Keverne, E. B. (2004). Brain evolution, chemosensory processing, and behavior. *Nutr Rev*, **62**, S218–223; discussion S224–41.

Keverne, E. B., and Curley, J. P. (2004). Vasopressin, oxytocin and social behavior. *Curr Opinion in Neurobiology*, **14**, 777–783.

King, B. C., and Nicholson, R. C. (2007) Advances in understanding CRH gene expression. *Frontiers in Bioscience*, **12**, 581–590.

Kirsch, P., Esslinger, C., Chen, Q., et al. (2005). Oxytocin modulates neural circuitry for social cognition and fear in humans. *J Neurosci*, **25**, 11489–11493.

Kluver, H. M., and Bucy, P. C. (1939). Preliminary analysis of functions of the temporal lobes in monkeys. *Archives of Neurology & Psychiatry*, **42**, 979–1000.

Koenigs M, Young L, et al., (2007). Damage to the prefrontal cortex increases utilitarian moral judgements. *Nature*, **446**, 7138, 908–911.

Koob, G. F., and LeMoal, M. (2001). Drug addiction, dysregulation of reward, and allostasis. *Neuropsychophamacology*, **24**, 94–129.

Koob, G. F., and LeMoal, M. (2005). *Neurobiology of addiction*. New York: Elsevier.

Korte, S. M., Koolhaas, J. M., Wingfield, J. C., and McEwen, B. S. (2004). The Darwinian concept of stress: Benefits of allostasis and costs of allostatic load and the trade-offs in health and disease. *Neuroscience and Biobehavioral Reviews*, **29**, 3–38.

Kosfeld, M., Heinrichs, M., Zak, P.J., Fischbacher, U. and Fehr E. (2005). Oxytocin increases trust in humans. *Science*, **435**, 673–676.

Krieckhaus, E. E., and Wolf, G. (1968). Interaction of innate mechanisms and latent learning. *Journal of Comparative and Physiological Psychology*, **65**, 197–201.

Kuhn, C. M., Pauk, J., and Schanberg, S. M. (1990). Endocrine responses to mother-infant separation in developing rats. *Dev Psychobiol*, **23**, 395–410.

Lakoff, G., and Johnson, M. (1999). *Philosophy in the Flesh*. New York: Basic Books.

Lashley, K. S. (1951). The problem of serial order in behavior. In L.A. Jeffress (Ed.), *Cerebral mechanisms in behavior*. Wiley, New York, 112–131.

LeDoux, J. E. (1995). *The emotional brain*. New York: Simon & Schuster.

Lehrman, (1958). Induction of broodiness by participation in courtship and nest-building in the ring dove. *J comp Physiol Psychol*, **51**, 32–36.

Lim, M. M., and Young, L. J. (2006). Neuropeptidergic regulation of affiliative behavior and social bonding in animals. *Hormones and Behavior*, **50**, 506–517.

Lind, R. W., Swanson, L. W., and Ganten, D. (1985). Organization of angiotensin II immunoreactive cells and fibers in the rat central nervous system. An immunohistochemical study. *Neuroendocrinology*, **40**, 2–24.

Liu, Y., and Wang, Z. X. (2003). Nucleus accumbens oxytocin and dopamine interact to regulate pair bond formation in female prairie voles. *Neuroscience*, **121**, 537–544.

Lovejoy, D. A., and Jahan, S. (2006). Phylogeny of corticotrophin-releasing factor family of peptides in the metazoan. *General & Comparative Endocrinology*, **146**, 1–8.

Lupien, S. J., Maheu, F., Tu, M., Fiocco, A., and Schramek, T. E. (2007). The effects of stress and stress hormones on human cognition: Implications for the field of brain and cognition. *Brain & Cognition*, **65**: 209–237.

Lyell, C. (1830–1833/1997). *Principles of geology*. New York: Penguin Books.

Maclean, P.D. (1990). *The triune brainin evolution*. New York: Plenium Press.

Maess, B., Koelsch, S., Gunter, T. C., and Friederici, A. D. (2001). Musical syntax is processed in Broca's area. *Nature Neuroscience*, **4** (5), 540–545.

Makino, S., Gold, P. W., and Schulkin, J. (1994 a/b). Corticosterone effects on corticotropin-releasing hormone mRNA in the central nucleus of the amygdala and the parvocellular region of the paraventricular nuclues of the hypothalamus and Effects of corticosterone on CRH mRNA and content in the bed nucleus of the amygdala and the paraventricular nucleus of the hypothalamus. *Brain Res*, **640**, 105–112, 141–149.

Malthus, T. R. (1798/1970). *An essay on the principle of population*. Baltimore, MD: Penguin Books.

Marler, C. A., Chu, J., and Wilczynski, W. (1995). Arginine vasotocin injection increases probability of calling in cricket frogs, but causes call changes characteristic of less aggressive males. *Horm Behav*, **29**, 554–570.

Marler, P. (1961). The logical analysis of animal communication. *J of Theoretical Biology*, **1**, 295–317.

Marler, P., and Hamilton, W. J. (1966). *Mechanisms of animal behavior*. New York: Wiley.

Marler, P., Peters, S., Ball, G. F., Duffy, A. M., Jr. and Wingfield, J. C. (1988). The role of sex steroids in the acquisition and production of birdsong. *Nature*, **336**, 770–772.

Martin, A., Wiggs, C. L., Ungerleider, L. G., and Haxby, J. V. (1996). Neural correlates of category specific knowledge. *Nature*, **379**, 649–652.

Mayr, E. (1963). *Animal species and evolution*. Cambridge, MA: Harvard University Press.

Mayr, E. (1991). *One long argument.* Cambridge, MA: Harvard University Press.

McCarthy, M. M. (2008). Estradiol and the developing brain. *Physiol Rev,* **88,** 91–134.

McEwen, B. S. (1995). Steroid actions on neuronal signalling. *Ernst Schering Research Foundation Lecture Series,* **27,** 1–45.

McEwen, B. S. (1998). Protective and damaging effects of stress mediators. *New Engl J Med,* **338,** 171–179.

McEwen, B. S. (2006). Sleep deprivation as a neurobiologic and physiologic stressor: Allostasis and allostatic load. *Metabolism,* **55,** S20-S23.

McEwen, B. S. and Wingfield, J. C. (2003). The concept of allostasis in biology and biomedicine. *Hormones and Behavior,* **43,** 2–15.

McGaugh, J. L. (2000). Memory – A century of consolidation. *Science,* **287,** 248–251.

McHenry, H. M. (2009). Human evolution. In M. Ruse and J. Travis (Eds), *Evolution: The First Four Billion Years.* Harvard University Press, Cambridge, Massachusetts, 256–280.

Meaney, M. J. (2001). Maternal care, gene expression, and the transmission of individual differences in stress reactivity across generations. *Annu Rev Neurosci,* **24,** 1161–1192.

Mellars, P. (2006). Going east: new genetic and archaeological perspectives on the modern human colonization of Eurasia. *Science,* **313,** 796–800.

Meltzoff, A. N. and Moore, M. K. (1977). Imitation of facial and manual gestures by human neonates. *Science,* **198,** 75–78.

Merali, Z., Anisman, H., James, J.S., Kent, P., and Schulkin, J. (2008). Effects of corticosterone on CRH and gastrin-releasing peptide release in response to an aversive stimulus in two regions of the forebrain. *Eur J of Neurosci,* **28,** 165–172.

Mercer, J.G., Lawrence, C.B., and Atkinson, T. (1996) Hypothalamic NPY and CRF gene-expression in the food-deprived Syrian hamster. *Physiology and Behavior,* **60:** 121–127.

Milgram, S. (1974), *Obedience to authority; an experimental view.* London: HarperCollins.

Mithen, S. (1996). *The prehistory of the mind.* London: Thames and Hudson.

Mithen, S. (2006) *The singing neanderthals: the origins of music, language, mind, and body.* Cambridge, MA: Harvard University Press.

Mitra, R., Jadhav, S., McEwen, B. S., Vyas, A., and Chattarji, S. (2005). Stress duration modulates the spatiotemporal patterns of spine formation in the basolateral amygdala. *PNAS,* **102,** 9371–9376.

Moll, J., and Schulkin, J. (2009). Social attachment and aversion in human moral cognition. *Neuroscience and biobehavioral reviews,* **33**(3), 456–465

Moore, F. L., and Orchinik, M. (1991). Multiple molecular action for steroids in the regulation of reproductive behaviors. *Sem in the Neurosci,* **3,** 489–496.

Moore, F. L., and Rose, J. D. (2002). Sensorimotor processing model. In D. W. Pfaff, *et al.,* (Eds.), *Hormones, brain and behavior.* New York: Academic Press.

Moore-Ede, M. C. (1986). Physiology of the circadian timing system: Predictive versus reactive homeostasis. *Am J Physiol,* **250,** R737–R752.

Moreno, J. (1995). *Deciding together: bioethics and moral consensus.* Oxford: Oxford University Press.

Morgan, M., and LeDoux, J. E. (1995). Differential contribution of dorsal and ventral medial prefrontal cortex to the acquisition and extinction of conditioned fear. *Behavioral Neuroscience,* **109** 681–688.

Morris, J. S., Frith, C. D., Perrett, D. I., *et al.* (1996). A differential neural response in the human amygdala to fearful and happy facial expressions. *Nature,* **383,** 812–815.

Mota, M. T., and Sousa, M. B. C. (2000). Prolactin levels of fathers and helpers related to alloparental care in common marmosets, Callithrix jacchus. *Folia Primatology*, **71**, 22–26.

Mrosovsky, N. (1990). *Rheostasis: The physiology of change*. New York: Oxford University Press.

Myers, D. A., Gibson, M., Schulkin, J., and Greenwood-Van-Meerveld, B. (2005). Corticosterone implants to the amygdala and type 1 CRH receptor regulation: Effects on behavior and colonic sensitivity. *Behavioural Brain Research*, **161**, 39–44.

Nauta, W. J. H. (1972). The central visceromotor system: A general survey. In C. H. Hockman (Ed.), *Limbic system mechanisms and autonomic function*. Springfield, IL: Charles C. Thomas.

Nelson, E., and Panksepp, J. (1998). Brain substrates of infant-mother attachment: Contributions of opioids, oxytocin, and norepinepherine. *Neuroscience and Biobehavioral Reviews*, **22**, 437–452.

Nemeroff, C. B., Krishnan, K. R., Reed, D., Leder, R., Beam, C., and Dunnick, N. R. (1992). Adrenal gland enlargement in major depression: A computed tomographic study. *Arch Gen Psychiatry*, **49**, 384–387.

Nephew, B. C., and Bridges, R. S. (2008). Arginine vasopressin V1a receptor antagonist impairs maternal memory in rats. *Physiology and Behavior*, **95**, 182–186.

Neuman, I. D. (2008). Brain oxytocin: A key regulator of emotional and social behaviours in oth females and males. *Journal of Neuroendocrinology*, **20**, 858–865.

Newman, S. W. (2002). Pheromone signaling access the medial extended amygdala. In D. W. Pfaff *et al.* (Eds.), *Hormones, brain and behavior*. New York: Academic Press.

Nicholson, R. C., King, B. R., and Smith R. (2004). Complex regulatory interactions control CRH gene expression. *Frontiers in Bioscience*, **9**, 32–39.

Norgren, R. (1995). Gustatory system. In *The rat nervous system*, 2nd edn. San Diego, CA: Academic Press, 751–771.

Nottebohm, F. (1994). The song circuits of the avian brain as a model system in which to study vocal learning, communication and manipulation. *Discussions in Neurosciences*, **10**, 72–81.

O'Doherty, J., Dayan, P., Schultz, J., Deichmann, R., Friston, K., and Dolan, R. J. (2004). Dissociable roles of ventral and dorsal striatum in instrumental conditioning. *Science*, **304**, 452–455.

Oftedal, O. T. (2002). The mammary gland and its origin during synapsid evolution. *J of Mamm Gland Biol and Neoplasia*, **7**, 225–252.

Olazabal, D. E., and Young, L. J. (2006). Species and individual differences in juvenile female alloparental care are associated with oxytocin receptor density in the striatum and the lateral septum. *Hormones and Behavior*, **49**, 681–687.

Pecina, S., Schulkin, J., and Berridge, K. C. (2006). Nucleus accumbens corticotropin-releasing factor increases cue-triggered motivation for sucrose reward: Paradoxical positive incentive effects in stress? *BMC Biol*, **4**, 8.

Petrovic, P., Kalisch, R., Singer, T., and Dolan, R. J. (2008). Oxytocin attenuates affective evaluations of conditioned faces and amygdala activity. *J Neurosci*, **28**, 6607–6615.

Pfaff, D. W. (1980). *Estrogens and brain function*. New York: Springer-Verlag.

Pfaff, D. W. (1999). *Drive*. Cambridge, MA: MIT Press.

Phillips, M. L, Young, A. W., Scott, S.K., *et al.* (1998). Neural responses to facial and vocal expressions of fear and disgust. *Proceedings of the Royal Society*, **265**, 1809–1817.

Pinker, S. (1994). *The language instinct*. New York: William Morrow and Co.

Porges, S. W. (1995). Orienting in a defensive world. *Psychophysiology*, **32**, 301–318.

Porges, S. W. (2003). Social engagement and attachment: a phylogenetic perspective. *Ann New York Acad Sci*, **1008**, 31–47.

Power, M. L., and Schulkin, J. (2006). Functions of CRH in anthropoid primates: From brain to placenta. *Am J of Human Biology*, **18**, 431–447.

Power, M. L. and Schulkin, J. (2009). *Evolution of obesity*. Baltimore, MD: Johns Hopkins University Press.

Powley, T.L. (1977). The ventromedial hypothalamic syndrome, satiety and cephalic phase. *Psychological Review*, **84**, 89–126.

Powley, T. L. (2000). Vagal circuitry mediating cephalic-phase responses to food. *Appetite*, **34**, 184–188.

Quirk, G. J., Russo, G. K., Barron, J. L., and Lebron, K. (2000). The role of ventromedial prefrontal cortex in the recovery of extinguished fear. *J Neurosci*, **20**, 6225–6231.

Radley, J. J., Arias, C. M., and Sawchenko, P. E. (2006). Regional differentiation of the medial prefrontal cortex in regulating adaptive responses to acute emotional stress. *Journal of Neuroscience*, **26**, 12967–12976.

Rakic. P. (2002). Evolving concepts of cortical radial and areal specification, *Prog Brain Res*, **136**, 265–280.

Richter, C. P. (1943). Total self-regulatory functions in animals and human beings. *Harvey Lectures*, **38**, 367–371.

Rimmele, U., Hediger, K., Heinrichs, M., and Klaver, P. (2009). Oxytocin makes a face in memory familiar. *The Journal of Neuroscience*, **29**, 38–42.

Rizzolatti, G., and Luppino G. (2001). The cortical motor system. *Neuron*, **31**, 889–901.

Roberts, R. L., Jenkins, K. T., Lawler, T., *et al.* (2001). Prolactin levels are elevated after infant carrying in parentally inexperienced common marmosets. *Physiol Behav*, **72**, 713–720.

Robson, S., and Wood, B. (2008). Hominin life history: Reconstruction and evolution. *J Anat*, **212**, 394–425.

Roitman M. F., Na, E., Anderson, G., *et al.* (2002). Induction of a salt appetite alters dendritic morphology in nucleus accumbens and sensitizes rats to amphetamine. *The Journal of Neuroscience*, **22**, 1–5.

Rolls, E. T. (2000). The orbitofrontal cortex and reward. *Cerebral Cortex*, **10**, 284–294.

Romero, L. M., Dickens, M. J., and Cyr, N. E. (2009). The reactive scope model: a new model integrating homeostasis, allostasis, and stress. *Hormones and Behavior*, **55**, 375–389.

Rosen, J. B. (2005). The neurobiology of conditioned and unconditioned fear: a neurobehavioral system analysis of the amygdala. *Behavioral and Cognitive Neuroscience Reviews*, **3**, 23–41.

Rosen, J. B., and Schulkin, J. (1998). From normal fear to pathalogical anxiety. *Psychol Rev*, **105**, 325–350.

Rozin, P. (1976). The evolution of intelligence and access to the cognitive unconscious. In J. Sprague and A. N. Epstein (Eds.), *Progress in psychobiology and physiological psychology*. New York: Academic Press.

Ruscio, M. G., Sweeny, T., Hazelton, J., Suppatkul, P., and Carter, C. S. (2007). Social environment regulates corticotropin releasing factor, corticosterone and vasopressin in juvenile prairie voles. *Hormones & Behavior*, **51**, 54–61.

Ryff, C. D., Singer, B. H., and Dienberg Love G. (2004). Positive health: connecting well-being with biology. *Phil R Soc London B*, **359**, 1386–1394.

Sabini, J., and Silver, M. (1982). *Moralities of everyday life*. Oxford: Oxford University Press.

Sanford, L. D., Nassar, P., Ross, R. J., Schulkin J., and Morrison, A. R. (1998). Prolactin Microinjections into the Amygdalar Central Nucleus Lead to Decreased NREM Sleep. *Sleep Research Online*, **1**, 109–113.

Saper, C. B. (1995). Central autonomic system. In G. Paxinos (Ed.), *The rat nervous system*. New York: Academic Press.

Sapolsky, R. M. (1992). *Stress: The aging brain and the mechanisms of neuron death*. Cambridge, MA: MIT Press.

Sapolsky, R. M. (1995). Social subordinance as a marker of hypercortisolism. Some unexpected subtleties. *Ann NY Acad Sci*, **771**, 626–639.

Savaskan, E., Ehrhardt, R., Schulz, A., Walter, M., and Schächinger, H. (2008). Post-learning intranasal oxytocin modulates human memory for facial identity. *Psychoneuroendocrinology*, **33**, 368–374.

Sawchenko, P. E. (1987). Evidence for local site of action for glucocorticoids in inhibiting CRH and vasopressin expression in the paraventricular nucleus. *Brain Res*, **17**, 213–223.

Sawchenko, P. E., Swanson, L. W., and Vale, W. W. (1984.) CRF: co-expression within distinct subset of oxytoci-vasopressin and neurotensin immunorective neurons in the hypothalamus of the male rat. *J. of Neuroscience*, **4**, 1118–1129.

Shachar-Dadon A, Schulkin, J, and Leshem M. (2009). Adversity before conception will affect adult progeny in rats. *Developmental Psychology*, **45**(1), 9–16.

Schmidt, L. A., Fox, N. A., Schulkin, J. and Gold, P. W. (1999). Behavioral and psychophysiological correlates of self-presentation in temperamentally shy children. *Developmental Psychobiology*, **35**, 119–135.

Schmidt, L. A., Fox, N. A., and Hamer, D. H. (2007). Evidence for a gene-gene interaction in children's behavior problems: association of 5-HTT short and DRD4 long genotypes with internalizing and externalizing behaviors in typically developing seven year-olds. *Developmental and Psychopathology*, **19**, 1105–1116.

Schulkin, J. (1991). *Sodium hunger: the search for a salty taste*. Cambridge: Cambridge University Press.

Schulkin, J. (1999). *The neuroendocrine regulation of behavior*. Cambridge, MA: MIT Press.

Schulkin, J. (2003). *Rethinking homeostasis*. Cambridge, MA: MIT Press.

Schulkin, J., Morgan, M. A., and Rosen, J. B. (2005). A neuroendocrine mechanism for sustaining fear. *Trends in Neuroscience*, **28**, 629–35.

Schultz, W. (2007). Multiple dopamine fujnctions at different time courses. *Ann. Rev. Neurosci*. **30**, 59–88.

Schwabl, H. (1993). Yolk is a source of maternal testosterone for developing birds. *Proc Natl Acad Sci USA*, **90**, 11446–11450.

Schwartz, C. E., Wright, C. I., Shin, L. M., Kagan, J., and Rauch, S. L. (2003). Inhibited and unhibited infants "grown up": adult amygdalar response to novelty. *Science*, **300**, 1952–1955.

Selye, H. (1956). *The stress of life*. New York: McGraw-Hill.

Shepard, J. D., Barron, K. W., and Myers, D. A. (2000). Corticosterone delivery to the amygdala increases corticotropin-releasing hormone mRNA in the central nucleus of the amygdala and anxiety-like behavior. *Brain Research*, **851**, 288–295.

Shepard, J. D., Liu, Y., Sassone-Corsi, P., and Aguilera, G. (2005). Role of glucocorticoids and cAMP-mediated repression in limiting corticotrophin-releasing hormone transcription during stress. *Journal of Neuroscience*, **25**, 4073–4081.

Silk, J. B. (2007). The adaptive value of sociality in mammalian groups. *Proc. Trans. R. Soc.*, **362**, 539–559.

Simpson, G. G. (1944). *Tempo and mode in evolution*. New York: Columbia University Press.

Singer, B., and Ryff, C. D. (1999). Hierarchies of life histories and associated health risks. *Ann N Y Acad Sci*, **896**, 96–115.

Singer, T., Snozzi, R., Bird, G., et al. (2008). Effects of oxytocin and prosocial behavior on brain responses to direct and vicariously experienced pain. *Emotion*, **8**, 781–791.

Smith, G. P. (1997). *Satiation from gut to brain*. Oxford: Oxford University Press.

Smith, R. (2007). Mechanisms of disease: parturition. *New England Journal of Medicine*, **356**, 271–283.

Smoller, J. W., Yamaki, L. H., Fagerness, J. A., et al., (2005) The corticotropin-releasing hormone gene and behavioral inhibition in children at risk for panic disorder. *Biol Psychiatry*, **57**, 1485–1492.

Stellar, E. (1954). The physiology of motivation. *Psychological review*, **61**, 5–22.

Stellar, J. R., and Stellar, E. (1985). *The neurobiology of motivation and reward*. New York: Springer-Verlag.

Sterling, P. (2004). Principles of allostasis: Optimal design, predictive regulation, psychopathology and rational therapeutics. In J. Schulkin (Ed.) *Allostasis, homeostasis and the costs of physiological adaptation*. Cambridge: Cambridge University Press.

Sterling, P., and Eyer, J. (1988). Allostasis: a new paradigm to explain arousal pathology. In S. Fisher and J. Reason (Eds.), *Handbook of life stress, cognition, and health*. New York: John Wiley & Sons.

Strand, F. L. (1999). *Neuropeptides: regulators of physiological processes*. Cambridge, MA: MIT Press.

Sullivan, R. M. and Gratton, A. (1999). Lateralized effects of medial prefrontal cortex lesions on neuroendocrine and autonomic stress responses in rats. *The Journal of Neuroscience*, **19**(7), 2834–2840.

Sumners, C., Gault, T.R., and Fregly, M.J. (1991). Potentiation of angiotensin II-induced drinking by gluocorticoids is a specific glucocorticoid Type II receptor (GR)-mediated event. *Brain Res*, **552**, 2–8.

Swanson, L.W. (2000). Cerebral hemisphere regulation of motivated behavior. *Brain Res*, **886**: 113–164.

Swanson, L. W. (1988). The neural basis of motivated behavior. *Acta Morphol Neurol Scand*, **26**, 165–176.

Swanson, L. W., and Simmons, D. M. (1989). Differential steroid hormone and neural influences on peptide mRNA levels in CRH cells of the paraventricular nucleus: a hybridization histochemical study in the rat. *J Comp Neurol*, **285**, 413–435.

Takahashi, L. K., Nakashima, B. R., Hong, H., and Watanabe, K. (2005). The smell of danger: a behavioral and neural analysis of predator odor-induced fear. *Neuroscience and Biobehavioral Reviews*, **29**, 1157–1167.

Takahashi, L. K., Turner, J. G., and Kalin, N. H. (1998). Prolonged stress-induced elevation in plasma corticosterone during pregnancy in the rat: implications for prenatal stress studies. *Psychoneuroendocrinology*, **23**, 571–581.

Taylor, S. E. (2006). Tend and befriend: biobehavioral bases of affiliation under stress. *Current Directions in Psychological Science*, **15**(6), 273–277.

Tessitore, A., Hariri, A. R., Fera, F., et al. (2002). Dopamine modulates the response of the human amygdala: a study in Parkinson's disease. *J. of Neuroscience*, **22**, 9099–9103.

Thompson, R. R., George, K., Walton, J. C., Orr, S. P., and Benson, J. (2006). Sex-specific influences of vasopressin on human social communication. *Proc Natl Acad Sci*, **103**, 7889–7894.

Thompson, B. L., Erickson, K. Schulkin, J., and Rosen, J. B. (2004). Corticosterone facilitates retention of contextual fear conditioning and increases CRH mRNA expression in the amygdala. *Behavioral Brain Research*, **149**, 209–215.

Tinbergen, N. (1951/1969). *The study of instinct*. Oxford: Oxford University Press.

Tomasello, M. (1999). *The cultural origins of human cognition*. Cambridge, MA: Harvard University Press.

Tomasello, M., Kruger, A. C., and Ratner, H. H. (1993). Cultural learning. *Behavioral and Brain Sciences*, **16**: 495–552.

Ullman, M. T. (2001). A neurocognitive perspective on language: the declarative procedural model. *Nature Neuroscience*, **9**, 266–286.

Ullman, M. T. (2004). Is Broca's area part of a basal ganglia thalamocortical circuit? *Cognition*, **92**, 231–270.

Unkelbach, C., Guastella, A.J., and Forgas, J. P. (2008) Oxytoctin selectively facilitates recognition of positive sex and relationship words. *Psychological Science* **19**, 1092–1094.

Urnäs-Mobey, K. (1998). Oxytocin may mediate the benefits of positive social interaction and emotions. *Psychoneuroendocrinology*, **23** (8), 819–835.

Vale, W., Spiess, J., Rivier, C., and Rivier, J. (1981). Characterization of a 41-residue ovine hypothalamic peptide that stimulates the secretion of corticotropin releasing hormone and beta-endorphin. *Science*, **213**, 1394–1397.

Valenstein, E. S. (2005). *The war of the soups and the sparks*. New York: Columbia University Press.

Voorhuis, T. A., De Kloet, E. R., and De Wied, D. (1991). Effect of a vasotocin analog on singing behavior in the canary. *Horm Behav*, **25**, 549–559.

Wagner, U., Wahle, M., Moritz, F., Wagner, U., Häntzschel, H., and Baerwald, C. G. (2006). Promoter polymorphisms regulating corticotrophin-releasing hormone transcription in vitro. *Horm Metab Res*, **38**, 69–75.

Wallis, O. C., Mac-Kwashie, A. O., Makri, G., and Wallis, M. (2005). Molecular evolution of prolactin in primates. *J Mol Evol*, **60**, 606–614.

Wang, Y., Herrmann, C. S., Moess, B., *et al.* (2000). Localization of early syntactic processes in frontal and temporal cortical areas: a magnetoencphalographic study. *Human Brain Mapping*, **11**, 1–11.

Warren, W. C., Hillier, L. W., Marshall Graves, J. A., *et al.* (2008). Genome analysis of the platypus reveals unique signatures of evolution. *Nature*, **455**, 256.

Watts, A. G., and Sanchez-Watts, G. (1995). Region-specific regulation of neuropeptide mRNAs in rat limbic forebrain neurones by aldosterone and corticosterone. *J Physiol (Lond)*, **484**, 721–736.

Weaver, I. C., Cervoni, N., Champagne, F. A., *et al.* (2004). Epigenetic programming by maternal behavior. *Nat Neurosci*, **7**, 847–854.

Wehr, T. A., Moul, D. E., Barbato, G., *et al.* (1993). Conservation of photoperiod-responsive mechanisms in humans. *Am J Physiol*, **265**, 846–857.

Windle, R. J., Shanks, N., Lightman, S. L., and Ingram, C. D. (1997). Central oxytocin administration reduces stress-induced corticosterone relase and anxiety behavior in rats. *Endocrinology*, **138**, 2829–2834.

Wingfield, J. C. (2004). Allostatic load and life cycles: Implication for neuroendocrine control mechanisms. In J. Schulkin (Ed.), *Allostasis, homeostasis and the costs of physiological adaptation*. Cambridge: Cambridge University Press.

Winslow, J. T., Hastings, N., Carter, C. S., Harbaugh, C. R., and Insel, T. R. (1993). A role for central vasopressin in pair bonding in monogamous prairie voles. *Nature*, **365**, 545–548.

Wolf, G. (1964). Sodium appetite elicited by aldosterone. *Psychonomic Sciences*, **1**, 211–212.

Wood, B. (1992). Origin and evolution of the genus homo. *Nature*, **355**, 783–790.

Woods, S. C., Hutton, R. A., and Makous, W. (1970). Conditioned insulin secretion in the albino rat. *Proceedings of the Society of Experimental Biology and Medicine*, **133**, 965–968.

Yao, M., Schulkin, J., and Denver. R. J. (2008). Evolutionary conserved glucocorticoid regulation of CRH. *Endorcinology*, **149**, 2352–2360.

Young, L. J., Winslow, J. T., Want, Z, *et al.* (1997). Gene targeting approaches to neuroendocrinology. *Hormones and Behavior*, **31**, 221–231.

Young, L. J., Nilsen, R., Waymire, K. G., MacGregor, G. R., and Insel, T. R. (1999). Increased affiliative response to vasopressin in mice expressing the via receptor from a manogamous vole. *Nature*, **400**, 766–768.

Zald, D. H., and Rauch, S. L. (2006). *The Orbitofrontal Cortex*. Oxford: Oxford University press.

Zimmer, C. (2005). *Smithsonian intimate guide to human origins*. Washington, DC: Smithsonian Books.

Index

193

Printed in the United States
by Baker & Taylor Publisher Services